Biochemistry Research Trends

Biochemistry Research Trends

Mineral Water: From Basic Research to Clinical Applications
Maria João Martins, PhD (Editor)
2022. ISBN: 978-1-68507-458-6 (Hardcover)
2022. ISBN: 978-1-68507-541-5 (eBook)

Terpenes and Terpenoids: Sources, Applications and Biological Significance
Charles A. Davies (Editor)
2022. ISBN: 978-1-68507-559-0 (Hardcover)
2022. ISBN: 978-1-68507-595-8 (eBook)

Circadian Rhythms and Their Importance
Rajeshwar P. Sinha, PhD (Editor)
2022. ISBN: 978-1-68507-547-7 (Hardcover)
2022. ISBN: 978-1-68507-585-9 (eBook)

A Biochemical View of Antioxidants
David Aebisher PhD, DSc
and Dorota Bartusik-Aebisher, PhD (Editors)
2021. ISBN: 978-1-68507-151-6 (Hardcover)
2021. ISBN: 978-1-68507-295-7 (eBook)

Volatile Oils: Production, Composition and Uses
Sunita Singh, PhD (Editor)
2021. ISBN: 978-1-68507-186-8 (Hardcover)
2021. ISBN: 978-1-68507-241-4 (eBook)

An Introduction to Drug Carriers
Mohammad Ashrafuzzaman, D.Sc. (Editor)
2021. ISBN: 978-1-68507-148-6 (Hardcover)
2021. ISBN: 978-1-68507-157-8 (eBook)

More information about this series can be found at
https://novapublishers.com/product-category/series/biochemistry-research-trends/

David Aebisher, PhD
and Dorota Bartusik-Aebisher, PhD
Editors

The Biochemical Guide to Enzymes

Copyright © 2023 by Nova Science Publishers, Inc.
DOI: 10.52305/JMNV4489.

All rights reserved. No part of this book may be reproduced, stored in a retrieval system or transmitted in any form or by any means: electronic, electrostatic, magnetic, tape, mechanical photocopying, recording or otherwise without the written permission of the Publisher.

We have partnered with Copyright Clearance Center to make it easy for you to obtain permissions to reuse content from this publication. Simply navigate to this publication's page on Nova's website and locate the "Get Permission" button below the title description. This button is linked directly to the title's permission page on copyright.com. Alternatively, you can visit copyright.com and search by title, ISBN, or ISSN.

For further questions about using the service on copyright.com, please contact:
Copyright Clearance Center
Phone: +1-(978) 750-8400   Fax: +1-(978) 750-4470   E-mail: info@copyright.com.

NOTICE TO THE READER

The Publisher has taken reasonable care in the preparation of this book, but makes no expressed or implied warranty of any kind and assumes no responsibility for any errors or omissions. No liability is assumed for incidental or consequential damages in connection with or arising out of information contained in this book. The Publisher shall not be liable for any special, consequential, or exemplary damages resulting, in whole or in part, from the readers' use of, or reliance upon, this material. Any parts of this book based on government reports are so indicated and copyright is claimed for those parts to the extent applicable to compilations of such works.

Independent verification should be sought for any data, advice or recommendations contained in this book. In addition, no responsibility is assumed by the Publisher for any injury and/or damage to persons or property arising from any methods, products, instructions, ideas or otherwise contained in this publication.

This publication is designed to provide accurate and authoritative information with regard to the subject matter covered herein. It is sold with the clear understanding that the Publisher is not engaged in rendering legal or any other professional services. If legal or any other expert assistance is required, the services of a competent person should be sought. FROM A DECLARATION OF PARTICIPANTS JOINTLY ADOPTED BY A COMMITTEE OF THE AMERICAN BAR ASSOCIATION AND A COMMITTEE OF PUBLISHERS.

Additional color graphics may be available in the e-book version of this book.

Library of Congress Cataloging-in-Publication Data

ISBN: 979-8-88697-410-2

Published by Nova Science Publishers, Inc. † New York

Contents

Preface		ix
Chapter 1	**Pyruvate Dehydrogenase**	1
	Dominika Bąk, David Aebisher	
	and Dorota Bartusik-Aebisher	
Chapter 2	**Glyceraldehyde-3-Phosphate Dehydrogenase**	5
	Patrycja Dębska, David Aebisher	
	and Dorota Bartusik-Aebisher	
Chapter 3	**Luciferase**	9
	Magdalena Paśko, David Aebisher	
	and Dorota Bartusik-Aebisher	
Chapter 4	**Acetaldehyde Dehydrogenase**	13
	Łukasz Karaś, David Aebisher	
	and Dorota Bartusik-Aebisher	
Chapter 5	**The Chemical Action of Glyceraldehyde-3-Phosphate Dehydrogenase**	19
	Magdalena Koman, David Aebisher	
	and Dorota Bartusik-Aebisher	
Chapter 6	**Sucrose**	25
	Małgorzata Kraska, David Aebisher	
	and Dorota Bartusik-Aebisher	
Chapter 7	**Dehydrogenases**	31
	Julia Motyka, David Aebisher	
	and Dorota Bartusik-Aebisher	
Chapter 8	**Telomerase**	35
	Katarzyna Koszarska, David Aebisher	
	and Dorota Bartusik-Aebisher	

Contents

Chapter 9   **Oxoglutarate Dehydrogenase** 41
Karolina Lach, David Aebisher
and Dorota Bartusik-Aebisher

Chapter 10   **Maltase** 47
Aleksandra Kotlińska, David Aebisher
and Dorota Bartusik-Aebisher

Chapter 11   **Aminolevulinic Acid Synthase** 53
Anita Majda, David Aebisher
and Dorota Bartusik-Aebisher

Chapter 12   **Deiodinase** 57
Wiktoria Kaliszewska, David Aebisher
and Dorota Bartusik-Aebisher

Chapter 13   **Phenylalanine Hydroxylase** 61
Weronika Pawelic, David Aebisher
and Dorota Bartusik-Aebisher

Chapter 14   **Nitrogenase (Flavodoxin)** 65
Martyna Paśko, David Aebisher
and Dorota Bartusik-Aebisher

Chapter 15   **DMSO Reductase** 71
Kacper Wygonik, David Aebisher
and Dorota Bartusik-Aebisher

Chapter 16   **Choline Acetyltransferase** 77
Maria Słobodzian, David Aebisher
and Dorota Bartusik-Aebisher

Chapter 17   **Glutaredoxins** 83
Radosław Starzyk, David Aebisher
and Dorota Bartusik-Aebisher

Chapter 18   **Lactase** 89
Bartosz Ziobro, David Aebisher
and Dorota Bartusik-Aebisher

Chapter 19   **Neuron Specific Enolase (NSE)** 95
Klaudia Dynarowicz, David Aebisher
and Dorota Bartusik-Aebisher

Chapter 20	**Hyaluronidase** 99	
	Anna Zdziebło, David Aebisher	
	and Dorota Bartusik-Aebisher	
Chapter 21	**Beta-Galactosidase** 105	
	Mateusz Warzocha, David Aebisher	
	and Dorota Bartusik-Aebisher	
Chapter 22	**Photodynamic Therapy** 109	
	Adamo Federica, David Aebisher	
	and Dorota Bartusik-Aebisher	
Chapter 23	**Methods of Diagnostics of Ulcerative Colitis** 117	
	Agnieszka Przygórzewska, Iga Serafin,	
	Kacper Rogóż, Paweł Woźnicki,	
	David Aebisher and Dorota Bartusik-Aebishe	
Chapter 24	**Barley Enzymes** 123	
	Dominika Leś, David Aebisher	
	and Dorota Bartusik-Aebisher	
Index	 127	
Editors' Contact Information 133		

Preface

In this book, *The Biochemical Guide to Enzymes*, the authors tried to demonstrate the high degree of enzymes used in metabolomic processes. The mechanism of enzymes action is provided. This book presents the discovery, chemical structure and mechanism of action of enzymes. The metabolomic functions of enzymes in treatment and diagnostics are also described.

Chapter 1 - Cancer is the leading cause of death in the world, with nearly 10 million people dying from cancer each year. As a result, for many years, doctors, pharmacists and biochemists from around the world have been working intensively on more and more effective therapies in the treatment of neoplastic diseases.

Chapter 2 - Glyceraldehyde-3-phosphate dehydrogenase (GAPDH) is one of the enzymes from the oxidoreductase class. There is one functional gene in the human body that codes for GAPDH, which is located on the 12th chromosome.

Chapter 3 - Luciferases are classes of enzymes that catalyze the oxidation reactions of luciferin. An example of this class of enzymes is firefly luciferase, which is classified as Photinus-luciferin: oxygen-4-oxidoreductase. It takes part in the reaction of ATP hydrolysis and decarboxylation, the product of which is firefly luciferin, referred to as LH2.

Chapter 4 - Acetaldehyde dehydrogenase is a group of enzymes found in the human body with a CAS number (9028-91-5). They constitute only the greater part of the enzymes from the dehydrogenase group, which are encoded by genes such as: ALDH1A1, ALDH2 and ALDH5. However, the presence of this enzyme depending on the race is different, and so people of the Northeast or people with blue eyes and blonde hair may show a mutation in the gene for this enzyme, which makes its effectiveness much more reduced.

Chapter 5 - Glyceraldehyde-3-phosphate dehydrogenase is a protein classified as oxidoreductase, previously known only for its participation in the glycolysis process (it converts G3P into 1,3-bisphosphoglycerate). At the end of the 20th century, however, it was discovered that it is a multifunctional enzyme that is also responsible for iron transport, regulation of gene expression, cytoskeleton formation, protein phosphorylation, cell apoptosis and RNA stabilization.

Chapter 6 - Proteins play many important functions in the human body and amid the enormous amount of work they do, it is also worth looking at their enzymatic role in the reactions of decomposition of complex substances into simple substances.

Chapter 7 - Dehydrogenases are a group of proteins that act as a catalyst, which means that by influencing the reaction, they do not change the composition of the final mixture or the equilibrium constant, but only accelerate the equilibrium state of the system by lowering the activation energy value.

Chapter 8 - Telomerase is a large enzyme with the structure of a ribonucleoprotein complex (RNP), which is responsible for the aging of cells, and thus the entire organism. The function of this enzyme is similar to RNA-dependent DNA polymerase.

Chapter 9 - Proteins are essential for the proper functioning of the cell and the entire organism. Enzymes responsible for the fast and coordinated course of metabolic and biochemical changes, crucial for meeting the energy needs of cells, are of particular importance. Such proteins include oxaglutarate dehydrogenase.

Chapter 10 - Without the presence of numerous enzymes, digesting carbohydrates, one of the main components of our diet, would not be possible. Between monosaccharide molecules there are usually α-glycosidic bonds, and less frequently, as in the case of lactose β-glycosides, which enable the formation of complex carbohydrates, taking the form of long, sometimes branched chains.

Chapter 11 - ALA synthase, or aminolevulinic acid synthase (ALAS), is an enzyme that occurs naturally in the cells of our body, it is precisely present in the mitochondria. This protein is involved in the synthesis of porphyrin, which is part of the heme. Almost all living organisms need heme, which together with globin forms hemoglobin, an oxygen transporter.

Chapter 12 - Iodothyronine deiodinases are selenoproteins acting as enzymes, characterized by the presence of selenocysteine (Sec) in the active site. In the human body, they occur in three isoforms: type I deiodinase (D1), type II deiodinase (D2) and type III deiodinase (D3). These proteins differ in their half-lives - the longest is D3 (12 hours), D1 (8 hours), and D2 is only 45 minutes.

Chapter 13 - Phenylalanine hydroxylase is a liver enzyme protein that belongs to the class of hydrolases, which means that they break chemical bonds when exposed to water. The mentioned enzyme catalyses the reaction of conversion of one amino acid (L-phenylalanine) into another (L-tyrosine). More precisely, this reaction consists in the addition of the -OH group to the carbon at the 4-position of the tyrosine.

Chapter 14 - Nitrogenases (nitrases) are enzymatic complexes belonging to the group of metalloenzymes, found in archaea and bacteria - aerobic and anaerobic, autotrophs, heterotrophs and cyanobacteria. They require the presence of magnesium ions, less often vanadium. They are not found in any eukaryotic organisms.

Chapter 15 - Dimethyl sulfoxide reductases (DMSO) are a family of very old evolutionarily mononuclear molybdoenzymes. They can only be detected in prokaryotes (Bacteria and Archaea domains).

Chapter 16 - Choline acetyltransferase (ChAT) is an enzyme produced in the central nervous system. This protein is synthesized in the body of cholinergic nerve cells, from where it is then transported to the axonal endings. Most probably, this transport takes place on a fast and slow road.

Chapter 17 - Glutaredoxins (Grxs) are small intracellular thiol enzymes that belong to reductases. They belong to the Grx system consisting of glutaredoxin, glutathione, glutathione reductase and NADPH. Glutaredoxins were discovered by Arne Holmgren, a Swedish professor of biochemistry in 1976. They were isolated from mutant Escherichia coli bacteria.

Chapter 18 - Lactase (LPH-Lactase phlorozin hydrolase) is a protein with a molecular weight of about 160 kDa belonging to the β-galactosidase enzymes. LPH is found in the enterocyte cell membrane that forms the brush border of the small intestine. This is a characteristic feature of mammals, with the greatest amount of LPH being found in the middle of the small intestine. The gene encoding LPH is located on the long arm of the second chromosome (2q21).

Chapter 19 - Neuron Specific Enolase (NSE) is a metal-activated metalloenzyme. NSE is one of five isomers of the glycolytic enzyme enolase.

Chapter 20 - Hyaluronidase is a protein with enzymatic properties that catalyzes the degradation of hyaluronic acid. Thanks to its ability to bind water, it has strong moisturizing properties.

Chapter 21 - Beta-galactosidase is an enzyme that commonly occurs in the human organism. It is taking part in the digestion of lactose, which may sometimes cause disorders of this process. New research points to the involvement of beta-gal in the cell aging processes. There are many methods of beta-gal activity detection, one of which is fluorescent. As the biomarker, it has been used for ovarian carcinoma cells detection. It also takes part in colorectal tumor imaging. Therefore, in the future, it may apply for diagnosis. The application of this enzyme has also been found in the food industry, especially to crack bonding in complex sugars.

Chapter 22 - Photodynamic therapy (or PDT) undoubtedly represents one of the most innovative, modern and technologically advanced dermatological therapies currently available. Initially introduced for the treatment of pre-cancerous lesions and for skin tumors deriving from the epidermis (and therefore not melanocytic), photodynamic therapy has been used to treat an increasingly broad spectrum of skin conditions, both pathological and aesthetic (the so-called photorejuvenation). Based on the use of a sensitizing agent, PDT uses a particular light source to obtain its therapeutic and cosmetic effects. From light, therefore, the cause of these types of injuries and skin aging, science has developed a technique that, through light itself, brings health and beauty to the skin.

Chapter 23 - Ulcerative colitis (UC) is a chronic inflammatory disease of the colon, belonging to inflammatory bowel diseases. Its etiology is unknown. The first reports of a disease with ulceration of the mucosa of the large intestine, which was also the cause of death of patients, come from 1793. Instead, the term ulcerative colitis was used by the British physician and researcher Samuel Wilks in 1859. He described the case of a 42-year-old woman who died after several months of diarrhea with a fever that was difficult to reduce. In a postmortem examination, he showed that the cause of the symptoms was ulcerative colitis and the terminal ileum.

Chapter 24 - Enzymes are multi-molecular, mostly protein catalysts that improve specific chemical reactions by lowering their activation energy. Almost all chemical reactions related to the functioning of living organisms, including viruses, require the participation of enzymes to achieve sufficient efficiency.

Chapter 1

Pyruvate Dehydrogenase

Dominika Bąk, David Aebisher and Dorota Bartusik-Aebisher[*]

Medical College of the University of Rzeszów, Rzeszów, Poland

Abstract

Cancer is the leading cause of death in the world, with nearly 10 million people dying from cancer each year. As a result, for many years, doctors, pharmacists and biochemists from around the world have been working intensively on more and more effective therapies in the treatment of neoplastic diseases.

Keywords: pyruvate dehydrogenase, citric acid, Krebs cycle

The citric acid cycle or the Krebs cycle or the tricarboxylic acid cycle is a cyclic series of biochemical reactions. The mechanism of this cycle was investigated in the 1930s by Sir Hans Krebs. In 1937, the scientist presented the key elements of the cycle, for which he was awarded the Nobel Prize in 1953. The Krebs cycle is a series of reactions taking place in the mitochondria, as a result of which acetyl-CoA residues are oxidized to Co_2, and coenzymes are reduced. This cycle, as an important metabolic pathway enabling both the oxidation of organic compounds and the synthesis of biosynthetic substrates, must be strictly controlled. The decisive moment for the delivery of the substrate to the cycle is the pyruvate dehygrogenase complex (Feng et al.

---

[*] Corresponding Author's Email: dbartusikaebisher@ur.edu.pl.

In: The Biochemical Guide to Enzymes
Editors: David Aebisher and Dorota Bartusik-Aebisher
ISBN: 979-8-88697-410-2
© 2023 Nova Science Publishers, Inc.

2018). The pyruvate dehydrogenase complex (Figure 1) consists of three enzymes: pyruvate dehydrogenase - E1, dihydrolipoyl transacetylase - E2, dihydrolipoyl dehydrogenase - E3, connected by non-covalent bonds that irreversibly carboxylate pyruvate to acetylcocoenzyme A, binding to the acetarboxylic acid glycosylase. The molecular weight of the complex ranges from 4 to 10 million Daltons (Ferrarini et al. 2021).

Pyruvate dehydrogenase dissociated from the complex has a mass of about 90,000. This enzyme consists of two different polypeptide chains (α and β). The shape of pyruvate dehydrogenase in the electron microscope photos is the tetrahedron. Inhibition of pyruvate dehydrogenase (PDC) activity by specific kinase-mediated phosphorylation (PDK) is the cause of many metabolic disorders, including cancer. The PDC/PDK relationship is the subject of scientific research, and the conducted research focused on the PDC/PDK axis in cancer treatment has resulted in the development of a new generation of small molecule PDK inhibitors. Research into the paramount role of PDC in cellular energy metabolism and its regulation by PDK may help in a biochemical understanding of cancer and other diseases, which will facilitate their treatment. In addition to inhibiting the activity of the PDC complex, a pyruvate dehydrogenase deficiency is a medically important aspect. The following chapter will discuss the above issues that are important from a medical point of view (Han et al. 2020).

Transcription factors such as Myc, Wnt and factors induced by tissue oxygen deficiency in relation to the oxygen demand leading to hypoxia - HIF, acting alone or in combination, lead to an increase in the transcription of one or more isoforms of pyruvate dehydrogenase kinase in a cancer cell. The function of active PDK is to phosphorylate one or more serine residues on the E1α subunit of the pyruvate dehydrogenase complex, which leads to inhibition of its activity. In addition, the kinase causes pyruvate-driven oxidative phosphorylation (OXPHOS) (Han et al. 2020). The reduced amount of OXPHOS reduces the production of reactive oxygen species and the production of mitochondrial chemical signals related to the onset of apoptosis. It has also been found that, in some types of tumors, the Src oncogene phosphorylates tyrosine residues on the E1α subunit of the pyruvate dehydrogenase complex, regardless of changes in PDK expression. Inhibition of PDC and an increase in the amount of glucose transporter and glycolytic enzymes by HIF and transcription factors causes an increase in pyruvate, lactate and hydrogen ions inside the tumor, which are stabilized by positive feedback HIF. This metabolic dependence induces a high rate of glycolysis in neoplastic cells under both hypoxic and normal oxygenation conditions (oxygen glycolysis - Warburg effect). Accumulated lactate is also responsible for the inhibition of various immune processes, which facilitates the tumor metastasis (Palmieri et al. 2020).

All this information led to research focused on the PDC/PDK axis in cancer treatment and the development of a new generation of small molecule PDK inhibitors. Known PDK inhibitors act at one of 4 binding sites: the pyruvate binding site, the nucleotide binding site, the lipoamide binding site, and the allosteric site. The anti-tumor PDK inhibitor is mitaplatin, which contains two DCA (dichloroacetate) molecules bound to cisplatin and may act selectively on tumor mitochondria, which are characterized by altered glucose metabolism compared to normal cells. Mitaplatin, upon entering the cell interior, dissociates into cisplatin and two DCA molecules, which enables the occurrence of a double toxic effect: cisplatin against DNA and dichloroacetate, which inhibits PDK and reverses the Warburg effect (Palmieri et al. 2020).

Deficiency of the pyruvate dehydrogenase complex (PDC) is a metabolic disease associated with a disorder of carbohydrate metabolism that results in the body becoming depleted of energy. This disease can be caused by a deficiency in each of the three subunits that make up the complex. The most common, however, is the deficiency of the E1 subunit - pyruvate dehydrogenase. A much rarer disorder is the E2-subunit deficiency of dihydrolipoyl transcetylase, as only isolated cases of this defect have been reported. Deficiency of the E3-subunit of dihydrolipol dehydrogenase is distinguished from the others by the presence of additional symptoms, such as: increased levels of α-ketoglutarate and keto acids accumulating in maple syrup disease (Stacpoole et al. 2017).

The most common deficiency of the E1 subunit causes symptoms already in the newborn, which progress over time, and these symptoms include: developmental delay, periodically resolving ataxia, decreased muscle tone, abnormal eye movements, seizures, facial dysmorphia, microcephaly. The childhood form of the disease often manifests itself periodically, and the disease has many etiologies, so the prognosis is often difficult to predict.

In order to diagnose pyruvate dehydrogenase deficiency, biochemical tests are performed that show the excessive concentration of lactate and pyruvate in the serum with the simultaneous correct ratio of their concentrations. In order to confirm the diagnosis, the activity of pyruvate dehydrogenase subunits in skin fibroblasts, lymphocytes or muscles is additionally determined. The primary management of pyruvate dehydrogenase deficiency is dietary treatment. The doctor follows a ketogenic diet and limits the patient's consumption of monosaccharides (Stacpoole et al. 2017).

An exciting field of research is PDK inhibitors in the treatment of cancer. Unfortunately, many questions regarding the safety of these inhibitors remain unanswered. It is very important to conduct research that will determine the role, safety and effectiveness of PDK inhibitors in their use as a comprehensive therapy in combination with standard radiotherapy and chemotherapy. It is also important to investigate the effect of these inhibitors on the ability to induce dehydration neuropathy, and to see if this effect is synergistic with that of standard chemotherapeutic agents that also have neurotoxic potential. Despite many blind people, there is great hope that the new generation of small molecule regulators of the PDC/PDK axis will contribute to a better understanding and treatment of an ever-increasing number of diseases.

The clinical manifestations of pyruvate dehydrogenase deficiency are extensive. The hallmark of this disease is that it usually manifests itself in early childhood with characteristic nervous disorders. It is very important to conduct a lot of research to evaluate the promising therapies for this debilitating disease.

References

Feng Y, Xiong Y, Qiao T, Li X, Jia L, Han Y. Lactate dehydrogenase A: A key player in carcinogenesis and potential target in cancer therapy. *Cancer Med.* 2018 Dec;7(12): 6124-6136.

Ferrarini MG, Nisimura LM, Girard RMBM, Alencar MB, Fragoso MSI, Araújo-Silva CA, Veiga AA, Abud APR, Nardelli SC, Vommaro RC, Silber AM, France-Sagot M, Ávila AR. Dichloroacetate and Pyruvate Metabolism: Pyruvate Dehydrogenase Kinases as Targets Worth Investigating for Effective Therapy of Toxoplasmosis. *mSphere.* 2021 Jan 6;6(1):e01002-20.

Han Y, Sun W, Ren D, Zhang J, He Z, Fedorova J, Sun X, Han F, Li J. SIRT1 agonism modulates cardiac NLRP3 inflammasome through pyruvate dehydrogenase during ischemia and reperfusion. *Redox Biol.* 2020 Jul;34:101538.

Palmieri EM, Gonzalez-Cotto M, Baseler WA, Davies LC, Ghesquière B, Maio N, Rice CM, Rouault TA, Cassel T, Higashi RM, Lane AN, Fan TW, Wink DA, McVicar DW. Nitric oxide orchestrates metabolic rewiring in M1 macrophages by targeting aconitase 2 and pyruvate dehydrogenase. *Nat. Commun.* 2020 Feb 4;11(1):698.

Stacpoole PW. Therapeutic Targeting of the Pyruvate Dehydrogenase Complex/Pyruvate Dehydrogenase Kinase (PDC/PDK) Axis in Cancer. *J. Natl. Cancer Inst.* 2017 Nov 1;109(11).

Chapter 2

Glyceraldehyde-3-Phosphate Dehydrogenase

Patrycja Dębska, David Aebisher[*] and Dorota Bartusik-Aebisher
Medical College of the University of Rzeszów, Rzeszów, Poland

Abstract

Glyceraldehyde-3-phosphate dehydrogenase (GAPDH) is one of the enzymes from the oxidoreductase class. There is one functional gene in the human body that codes for GAPDH, which is located on the 12th chromosome.

Keywords: glyceraldehyde-3-phosphate, chromosome, cellular proteins

This enzyme is present in both eukaryotic and prokaryotic cells where it is present in large amounts - about 10-20% of the total content of cellular proteins. GAPDH is most abundant in the cytoplasm, but is also present in the nucleus and mitochondria. In mammals, G3P dehydrogenase exists as a tetramer consisting of four identical monomers. Its mass is approximately 143 kDa. GAPDH can also exist as a monomer - in the nucleus or a dimer - in the mitochondria (Kosova et al., 2017).

Glyceraldehyde-3-phosphate dehydrogenase is a multitasking protein because it can perform many different functions, not only related to glycolysis. The primary function of this enzyme is to catalyze the sixth stage of glycolysis, i.e., the oxidation of G3P to 1,3-bisphosphoglycerate. GAPDH is dependent

---

[*] Corresponding Author's Email: dbartusikaebisher@ur.edu.pl.

In: The Biochemical Guide to Enzymes
Editors: David Aebisher and Dorota Bartusik-Aebisher
ISBN: 979-8-88697-410-2
© 2023 Nova Science Publishers, Inc.

on NAD +. The catalytic ability of GAPDH is essential for the cells of all organisms to maintain adequate glycolytic flow. This glycolytic protein was also used as a model protein for the analysis of e.g., enzymatic mechanisms and protein structure. For a long time, this enzyme has been recognized as important in energy metabolism and in the synthesis of ATP and pyruvate through anaerobic glycolysis taking place in the cytoplasm. Dehydrogenase participates, inter alia, in cell membrane fusion, cytoskeleton formation, vesicular transport, and in maintaining DNA integrity (Kosova et al., 2017).

One of the most important functions of G3P dehydrogenase is participation in glycolysis. In the first stage of catalysis, the glyceraldehyde 3-phosphate reacts with the hydrosulfide group of the cysteine residue, which is present in the active center of the enzyme. As a result of this reaction, hemitioacetal is produced. In the next stage, the hydride ion is transferred from the intermediate product, hemitioacetal to the C4 carbon atom of the NAD + molecule. As a result of this reaction, NADH and the thioester are formed. Subsequently, NADH dissociates under the influence of the next NAD + molecule, while the thioester reacts with orthophosphate to form 1,3-bisphosphoglycerate. The effect of dimethyl fumarate on GAPDH and oxygen glycolysis was investigated. Dimethyl fumarate (DMF), a derivative of the Krebs intermediate cycle fumarate, is an immunomodulatory drug used to treat multiple sclerosis and psoriasis. Studies have shown that DMF succinates and inactivates the catalytic cysteine of the glycolytic enzyme glycerol 3-phosphate dehydrogenase (GAPDH) in mice and humans, both in vitro and in vivo. Thus, it down-regulates oxygen glycolysis in activated myeloid and lymphoid cells, which mediates its anti-inflammatory effect (Muronetz et al., 2017).

An equally important function of G3P dehydrogenase is to maintain DNA integrity. However, the mechanism of GAPDH translocation into the nucleus is not exactly known. GAPDH has been shown to interact with some types of DNA damage, such as apurine and apyrimidine sites, nucleotide analogues, and covalent DNA adducts with alkylating agents. Moreover, GAPDH may interact with proteins that are involved in DNA repair, e.g., APE1, PARP1, HMGB1 and HMGB2. The cell's ability to repair DNA is essential to the integrity of the entire genome as well as the integrity of genetic information. DNA is involved in the repair processes of, inter alia, the enzyme is uracil glycosylase (UNG). UDG-like activity is found in the nucleus of GAPDH. Glyceraldehyde-3-phosphate dehydrogenase plays a role in the regulation of telomere length. The end sections of the chromosomes are shortened with each cell division. The conducted studies have shown that glyceraldehyde-3-

phosphate dehydrogenase present in the nucleus binds to telomeric DNA, thus protecting it against shortening during replication. In addition, GAPDH has been shown to protect telomeres from degradation caused by anti-cancer drugs, including oxorubicin (Muronetz et al., 2019).

More and more studies indicate the participation of GAPDH in neurodegenerative diseases. This enzyme is one of the main components of neurofibrillary tangles (NFT), amyloid plaques and Levi bodies in postmortem brain tissues obtained from patients diagnosed with Alzheimer's and Parkinson's disease. It is likely that GAPDH accumulates in amyloid plaques and neurofibrillary tangles as a result of the oxidation and degradation of GAPDH. The result of this phenomenon is oxidative stress. Neuroproteomics revealed high-affinity interactions between GAPDH and proteins that are associated with Alzheimer's disease. These proteins include: β-amyloid, β-amyloid precursor protein, and tau. GAPDH also interacts with other proteins important in neurodegenerative diseases such as α-synuclein (Parkinson's disease) and huntingtin (Huntington's disease). The described interaction may lead to the impairment of the glycolytic function of GAPDH in Alzheimer's disease and may herald its function in apoptosis. Therefore, in neurodegenerative diseases, overexpression of GAPDH is observed with a concomitant reduction of the glycolytic activity of the enzyme. One of the reasons for the decreased activity of GAPDH is the modification of the cysteine residue located in the active center (Cys-152). GAPDH is involved in various nuclear and cytosolic pathways. It also contributes to the regulation of Ca^{2+} influx from the endoplasmic reticulum. GAPDH participates in retrograde transport between the tubular-alveolar intermediate structure (VTC) and the endoplasmic reticulum (ER) (Sirover et al., 2020).

Glyceraldehyde-3-phosphate dehydrogenase is a protein with many different functions. By catalyzing the oxidation of 3-phosphoglycerate to 1,3-bisphosphoglycerate during the glycolysis process, GAPDH is one of the factors enabling this process to take place. Most of these functions, however, are unrelated to involvement in the glycolytic pathway. GAPDH is ubiquitous and occurs in all areas of life. This enzyme plays an important role in many cellular and extracellular processes, including maintaining DNA integrity, regulating telomere length or participating in nuclear and cytosolic pathways, which makes it a very important enzyme for the proper functioning of the biochemical processes of both cells and the whole organism. This enzyme plays a role in many diseases including pathogenic, cardiovascular, degenerative, diabetic and oncogenic diseases. Demonstrating the correlation between G3D dehydrogenase and neurodegenerative diseases, such as

Alzheimer's disease or Parkinson's disease, allows for the possibility of further research on the activity of this enzyme in relation to the above diseases. Despite the many functions that this enzyme performs, and despite the different locations of these functions, glyceraldehyde-3-phosphate dehydrogenase in humans has only one functional gene. Some of the functions of GAPDH that have already been described in the literature require confirmation by experiment. Despite numerous studies already carried out, there is still no comprehensive knowledge about this extremely important enzyme (White et al., 2017).

References

Kosova A. A., Khodyreva S. N., Lavrik O. I. Role of Glyceraldehyde-3-Phosphate Dehydrogenase (GAPDH) in DNA Repair. *Biochemistry* (Mosc). 2017 Jun;82(6):643-654.

Muronetz V. I., Barinova K. V., Stroylova Y. Y., Semenyuk P. I., Schmalhausen E. V. Glyceraldehyde-3-phosphate dehydrogenase: Aggregation mechanisms and impact on amyloid neurodegenerative diseases. *Int J Biol Macromol.* 2017 Jul;100:55-66.

Muronetz V. I., Melnikova A. K., Barinova K. V., Schmalhausen E. V. Inhibitors of Glyceraldehyde 3-Phosphate Dehydrogenase and Unexpected Effects of Its Reduced Activity. *Biochemistry* (Mosc). 2019 Nov;84(11):1268-1279.

Sirover M. A. Moonlighting glyceraldehyde-3-phosphate dehydrogenase: posttranslational modification, protein and nucleic acid interactions in normal cells and in human pathology. *Crit Rev Biochem Mol Biol.* 2020 Aug;55(4):354-371.

White M. R., Garcin E. D. D-Glyceraldehyde-3-Phosphate Dehydrogenase Structure and Function. *Subcell Biochem.* 2017;83:413-453.

Chapter 3

Luciferase

Magdalena Paśko, David Aebisher and Dorota Bartusik-Aebisher[*]

Medical College of the University of Rzeszów, Rzeszów, Poland

Abstract

Luciferases are classes of enzymes that catalyze the oxidation reactions of luciferin. An example of this class of enzymes is firefly luciferase, which is classified as Photinus-luciferin: oxygen-4-oxidoreductase. It takes part in the reaction of ATP hydrolysis and decarboxylation, the product of which is firefly luciferin, referred to as LH2.

Keywords: luciferase, enzyme, biochemistry

Luciferase is a protein that forms two compact domains connected by a linker peptide that creates a large space between them. In an environment characterized by a pH of 7.5, the emission spectrum of luciferase falls in the yellow-green region (550-570 nm), but it is dependent on the temperature, pH and heavy metal cations to which the enzyme is sensitive (Zhou et al. 2018). This enzyme takes part in a bioluminescent reaction, which is a nucleophilic displacement of SN2, which includes the carboxyl group at the C4 carbon in the D-LH2 thiazoline ring and the phosphate groups of ATP. Coenzyme A (CoA) added to the luciferase reaction mixture can stabilize the emitted light, preventing the flash from occurring. On the other hand, its addition to a

---

[*] Corresponding Author's Email: dbartusikaebisher@ur.edu.pl.

In: The Biochemical Guide to Enzymes
Editors: David Aebisher and Dorota Bartusik-Aebisher
ISBN: 979-8-88697-410-2
© 2023 Nova Science Publishers, Inc.

mixture that has already produced light causes a secondary flash of light. Although the chemical structure and properties of LH2 are well known, the biosynthesis of this enzyme is relatively unknown. It is known that chemical synthesis of LH2 results in two enantiomers, D- and L, and that only the L-form of amino acids occurs naturally in peptides and proteins, therefore it has been found that D-LH2 can be produced from L-LH2 by enzymatically catalyzed inversion with CoA. Luciferases also have many important functions (Kim et al. 2018).

Each luciferase belongs to oxidoreductases, i.e., a class of enzymes that catalyze oxidation and reduction reactions. They transfer electrons and hydrogen atoms between the molecules of the oxidant and the reducing agent and the oxidant. These enzymes require molecular oxygen to react. Most of the luciferases tested have been detected in animals such as fireflies and many marine animals such as copepods, jellyfish and sea pansy. Luciferases are also found in glowing fungi such as the Jack-O-Lantern mushroom, and in glowing bacteria and dinoflagellates. Luciferase is folded into two domains linked by a flexible linker peptide. It is a small C-terminal domain and a large N-terminal domain. This protein takes a "closed form" during the formation of the high-energy intermediate that is responsible for light emission. This creates a hydrophobic pocket around the active site. A uniform mechanism of action of this enzyme cannot be established because it depends on the relationship between luciferin and luciferase, and luciferases are obtained from various proteins. The use of luciferase is very broad. For example, it serves as a rapporteur to assess transcriptional activity in cells transferred with a genetic construct containing the luciferase gene under the control of a selected promoter. Luciferase is also used to detect caspase and cytochrome P450 activity by using coupled or two-step assays (Lee et al. 2019).

This protein is also very sensitive to temperature changes, especially its increase, so it is often used to study the peptide denaturation process. In addition, using genetic engineering methods, luciferase genes can be synthesized and introduced into organisms or transferred into cells. Gene fusion uses luciferase reporter proteins and a fluorescent protein called GFP which is green in color. This technology allows you to visualize the promoter activity in detail. Optical cell tracking in small laboratory animals uses bioluminescent imaging (BLI). It enables the simultaneous visualization of the monitoring of the expression of two divergent luciferase proteins using their specific substrates through low background autofluorescence and cellular toxicity. Luciferase is involved in the light production reaction, forming adenyloxylcyphein on its surface. This results in a relatively short half-life of light emission with flash-type kinetics. The Firefly Luciferase HTS (SCT150) test prevents the formation of adenyloxylciferin on the luciferase surface by using a mixture of substances that convert a chemical reaction. Consequently, a long-lasting glow is created as a signal (Nair et al. 2018).

This kit allows the determination of a highly sensitive firefly luciferase reporter gene for the quantification of firefly luciferase expression in mammalian cells with a signal half-life of approximately 3 hours. Flash-type tests have a higher luminescence signal compared to luminescent-type luciferase tests. The accuracy of the assay depends on the level of luciferase expression in your experimental system and on the sensitivity of the luminometer. There are heterogeneous luciferase tests. Those that use firefly luciferase use luciferin in the presence of magnesium, oxygen, and ATP to produce 550-70 nM green or yellow light. In turn, Renilla Luciferase (SCT153) tests use coelenterazine and oxygen. They produce blue light with a length of 480 nM. For the in vivo and in vitro gene function analysis, Renilli luciferase is used as the reporter gene, since it does not require post-translational modification for its activity, and therefore it can act as a genetic reporter immediately after translation. In addition, light is also emitted by the Renilla luciferase substrate coelentrazine (Park et al. 2020).

Luciferase, which is an oxidoreductase, thanks to its structure (creating two compact domains connected by a linker peptide creating a large space between them) catalyzes oxidation and reduction reactions, which is why it is used for many purposes, such as studying protein denaturation, because it is sensitive to temperature changes. Luciferase is also used to detect caspase and cytochrome P450 activity and serves as a rapporteur to assess transcriptional activity in cells transferred with a genetic construct containing the luciferase gene. Moreover, luciferase is involved in the light production reaction, forming adenyloxylcyphene on its surface. S and by genetic engineering, luciferase genes can be synthesized and introduced into organisms or transferred into cells. The above applications of luciferase indicate the enormous potential of this enzyme in the future, which results from the specific property of luciferase, which is the fact that when it is added to a mixture that has already produced light, it causes a secondary flash of light. Also, many scientific studies are based on the specificity of this enzyme. The accuracy of the assay depends on the level of luciferase expression in your experimental system and on the sensitivity of the luminometer. There are heterogeneous luciferase tests, however, Renilla luciferase is used to analyze gene function *in vivo* and *in vitro*.

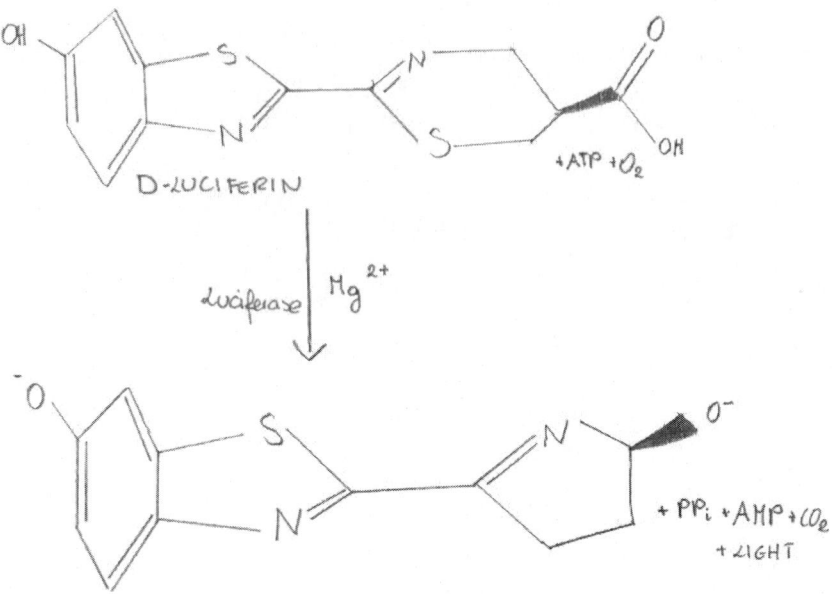

Figure 1. Luciferase catalyzed reaction.

References

Kim DS, Choi JR, Ko JA, Kim K. Re-engineering of Bacterial Luciferase; For New Aspects of Bioluminescence. *Curr. Protein Pept. Sci.* 2018;19(1):16-21.

Lee J, Müller F, Visser AJWG. The Sensitized Bioluminescence Mechanism of Bacterial Luciferase. *Photochem. Photobiol.* 2019 May;95(3):679-704.

Nair AK, Baier LJ. Using Luciferase Reporter Assays to Identify Functional Variants at Disease-Associated Loci. *Methods Mol. Biol.* 2018;1706:303-319.

Park SY, Song SH, Palmateer B, Pal A, Petersen ED, Shall GP, Welchko RM, Ibata K, Miyawaki A, Augustine GJ, Hochgeschwender U. Novel luciferase-opsin combinations for improved luminopsins. *J. Neurosci. Res.* 2020 Mar;98(3):410-421.

Zhou Z, Bi G, Zhou JM. Luciferase Complementation Assay for Protein-Protein Interactions in Plants. *Curr. Protoc. Plant Biol.* 2018 Mar;3(1):42-50.

Chapter 4

Acetaldehyde Dehydrogenase

Łukasz Karaś, David Aebisher and Dorota Bartusik-Aebisher[*]

Medical College of the University of Rzeszów, Rzeszów, Poland

Abstract

Acetaldehyde dehydrogenase is a group of enzymes found in the human body with a CAS number (9028-91-5). They constitute only the greater part of the enzymes from the dehydrogenase group, which are encoded by genes such as: ALDH1A1, ALDH2 and ALDH5. However, the presence of this enzyme depending on the race is different, and so people of the Northeast or people with blue eyes and blonde hair may show a mutation in the gene for this enzyme, which makes its effectiveness much more reduced.

Keywords: acetaldehyde dehydrogenase; enzyme; biochemistry

The above forementioned people, by consuming alcohol too regularly, are exposed to diseases such as liver damage, alcoholic asthma or cancers of the throat and even esophagus. One of the organs where we can find its presence is the liver, in which it takes part in the detoxification of ethanol alcohol by decomposing its oxidized form, i.e., acetaldehyde, into acetic acid. The role of this enzyme in this process is very important because acetic aldehyde is a compound that accumulates in such organs and is much more harmful than ethanol alcohol, so converting it into acetic acid is a very important process.

---

[*] Corresponding Author's Email: dbartusikaebisher@ur.edu.pl.

In: The Biochemical Guide to Enzymes
Editors: David Aebisher and Dorota Bartusik-Aebisher
ISBN: 979-8-88697-410-2
© 2023 Nova Science Publishers, Inc.

Recent studies show that this enzyme is not only involved in neutralizing acetaldehyde but is also involved in vitamin metabolism, and the efficient operation of this enzyme protects us against the so-called. "Visceral obesity" (Pang et al., 2017).

One of the amino acids that make up acetaldehyde dehydrogenase is cysteine-302, which is one of the 3 Cys residues that is crucial for the catalytic function of this enzyme. The rest is formed in the process of alkylation by iodoactamide in the cytosolic and mitochondrial isozymes. Additionally, it should be mentioned here that the sequence Gln-Gly-Gln-Cys preceding Cys-302 is conserved in both isozymes found in both human and equine. This fact can only confirm to us that Cys-302 is crucial for the catalytic functions of this enzyme. Thanks to research carried out in 1995 With site-directed mutagenase, glutamate-268 was found to be the major component of hepatic acetaldehyde dehydrogenase and critical for catalytic activity. The researchers found that the activity of the mutant enzymes could not be restored by adding general bases, they suggested that this residue functions as a rule of thumb for the activation of the essential Cys-302 residue (Wüthrich et al., 2018).

Work from 2003. She also proved that in bacteria, the acylating acetaldehyde dehydrogenase forms a difunctional heterodimer containing metal dependent 4-hydroxy-2-ketovalerate aldose. This enzyme in bacteria is used to degrade toxic aromatic compounds, and the crystalline structure of the enzyme may indicate that intermediates transferred directly between active sites via a culture-phobic intermediate channel provide a non-reactive environment where reactive acetaldehyde intermediate can be transferred from the aldose active site to the active site on acetaldehyde dehydrogenase. Such transport between proteins allows for efficient and energetically beneficial transport of substrates from the active site of the first enzyme to the active site of the second enzyme (Wakabayashi et al., 2019).

Acetaldehyde 2 dehydrogenase (ALDH2), as mentioned above, is a mitochondrial enzyme that detoxifies acetaldehyde but also endogenous lipid aldehydes, and a large part of the research suggests it's very important function in protection against cardiovascular diseases. According to statistics, about 40% of the population of people from the regions of Asia are carriers of the single nucleotide polymorphism ALDH2 rs671, and thus an increase in the incidence of CVD is noticeable. Unfortunately, at the present level of knowledge, the role of ALDH2 in the CVD prevention process is not well understood. In addition, it was found that mice with a doubled amount of ALDH2 / LDLR (DKO) have a clearly reduced number of atherosclerotic lesions compared to LDLR-KO mice, while in ALDH2 / APOE-DKO mice an increase in atherosclerotic lesions is visible. This statistic suggests to us that there is an interaction of ALDH2 with LDLR. Further research indicates to us that in the absence of LDLR, AMPK phosphorylates ALDH2 together with threonine 356 and allows its nuclear translocation. In contrast, nuclear ALDH2 interacts with HDAC3 and represses the transcription of the lysosomal proton pump protein ATP6V0E2, which is critical for the maintenance of lysosomal function, the degradation of low-density oxidized lipid proteins and their autophagy (Zhong et al., 2019).

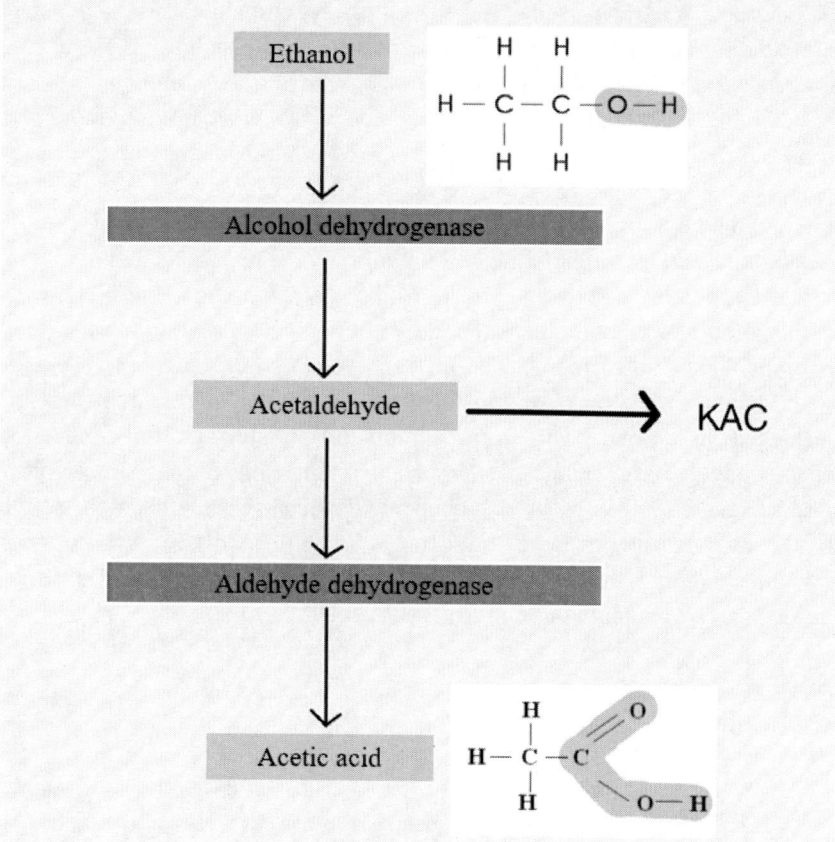

Figure 1. Diagram describing the distribution of ethanol in the human body.

Also, a very interesting conclusion should be mentioned here related to the interaction of the cytosolic C-terminus of LDLR with AMPK blocking ALDH2 phosphorylation and subsequent nuclear translocation, while the

ALDH2 rs671 mutation occurring in human macrophages significantly weakens this interaction, which in turn releases ALDH2 into the nucleus for the purpose of suppressing ATP6V0E2 expression. This reaction very quickly leads to an increase in the number of foam cells due to impaired lysosomal function occurring in the cell. The above study clearly shows us the very strong influence of the metabolic activity of ALDH2 on the breakdown of LDLR and its molecular mechanism by which in people without the rs671 SNP mutation, the incidence of CVD is much lower (Wang et al., 2020).

Acetaldehyde dehydrogenase is an enzyme classified as a dehydrogenase, i.e., a group of oxidoreductases, the main catalytic function of which consists in oxidation and reduction reactions with the use of protons and electrons. It is an enzyme widespread among mammals, occurring most often in the cytosol of cells, the endoplasmic reticulum, and above all, its greatest amount can be found in the mitochondrial matrix.Its redox functions, as can be seen in the materials above, are not fully understood and its phenotypic varieties caused by mutations within the genes coding for it can lead to many changes in the physiology of the living organism. In addition to its metabolic activity related to the detoxification of acetaldehyde, which is highly toxic to our body, which varies among races and makes the alcohol tolerance of people from different parts of the world different. The influence of this enzyme's activity on the work of our circulatory system is also noticeable, and as research shows, its influence on this area is very important, but unfortunately not fully understood at this stage of knowledge. A very clear relationship between the work of ALDH2 and LDLR shows us that deeper and extensive research on the activity of this enzyme may in the future bring us a number of new and innovative therapies helping us in the fight against atherosclerotic changes, which, unfortunately, are quite large in Poland. problem.

References

Pang J, Wang J, Zhang Y, Xu F, Chen Y. Targeting acetaldehyde dehydrogenase 2 (ALDH2) in heart failure-Recent insights and perspectives. *Biochim. Biophys. Acta Mol. Basis Dis.* 2017 Aug;1863(8):1933-1941.

Wakabayashi Y, Tamura Y, Kouzaki K, Kikuchi N, Hiranuma K, Menuki K, Tajima T, Yamanaka Y, Sakai A, Nakayama K I, Kawamoto T, Kitagawa K, Nakazato K. Acetaldehyde dehydrogenase 2 deficiency increases mitochondrial reactive oxygen species emission and induces mitochondrial protease Omi/HtrA2 in skeletal muscle. *Am. J. Physiol. Regul. Integr. Comp. Physiol.* 2020 Apr 1;318(4):R677-R690.

Wang W, Wang C, Xu H, Gao Y. Aldehyde Dehydrogenase, Liver Disease and Cancer. *Int. J. Biol. Sci.* 2020 Jan 22;16(6):921-934.

Wüthrich B. Allergic and intolerance reactions to wine. *Allergol. Select.* 2018 Sep. 1;2(1): 80-88.

Zhong S, Li L, Zhang YL, Zhang L, Lu J, Guo S, Liang N, Ge J, Zhu M, Tao Y, Wu YC, Yin H. Acetaldehyde dehydrogenase 2 interactions with LDLR and AMPK regulate foam cell formation. J Clin Invest. 2019 Jan 2;129(1):252-267.

Chapter 5

The Chemical Action of Glyceraldehyde-3-Phosphate Dehydrogenase

**Magdalena Koman, David Aebisher
and Dorota Bartusik-Aebisher***
Medical College of the University of Rzeszów, Rzeszów, Poland

Abstract

Glyceraldehyde-3-phosphate dehydrogenase is a protein classified as oxidoreductase, previously known only for its participation in the glycolysis process (it converts G3P into 1,3-bisphosphoglycerate). At the end of the 20th century, however, it was discovered that it is a multifunctional enzyme that is also responsible for iron transport, regulation of gene expression, cytoskeleton formation, protein phosphorylation, cell apoptosis and RNA stabilization.

Keywords: glyceraldehyde-3-phosphate, multifunctional enzyme, biochemistry

It was also found that the enzyme plays an important role in the pathophysiology of neurodegenerative diseases such as Parkinson's, Alzheimer's and Huntington's diseases. In this chapter, however, special attention is paid to the role of dehydrogenase in glycolysis and apoptosis (Kosova et al. 2017).

---

* Corresponding Author's Email: dbartusikaebisher@ur.edu.pl.

In: The Biochemical Guide to Enzymes
Editors: David Aebisher and Dorota Bartusik-Aebisher
ISBN: 979-8-88697-410-2
© 2023 Nova Science Publishers, Inc.

The above protein is encoded as a monomer by the 2597 gene, located on chromosome 12. The described enzyme, however, occurs naturally as a tetramer with a total atomic mass of approximately 143 kDa. Originally, this protein was isolated from human spleen. The percentage of glyceraldehyde-3-phosphate dehydrogenase among all cellular proteins is approx. 10-20%. Its most common place of occurrence in the cell is the cytoplasm. In smaller amounts, however, it is found in the mitochondria and the cell nucleus. Interestingly, in red blood cells it is located on the inner part of the cell membrane (Muronetz et al. 2019).

List of Important Abbreviations

- GAPDH: glyceraldehyde-3-phosphate dehydrogenase
- G3P: glyceraldehyde-3-phosphate
- 1,3-BPG: 1,3-bisphosphoglycerate
- NAD +: reduced form of nicotinamide adenine dinucleotide
- Siah-1: E3 ubiquitin ligase
- GOSPEL: (GADPH's competitor of siah protein enhances life) protein binding to GAPDH in competition with Siah-1 protein

GAPDH is an enzyme in the form of a tetramer, although less commonly it may be a monomer or a dimer. Each subunit is a 335 amino acid polypeptide with a molecular weight of 36 kDa. Monomers have 2 domains (Figure 1):

- N-terminal
- C-terminal

The first is responsible for binding NAD +. Its structure resembles similar domains in other dehydrogenases. It has Rossmann folds, thanks to which it can bind tubulin and the 3'UTR and 5'UTR ends of mRNA. The above-mentioned structures consist of centrally arranged, parallel to each other, the so-called β-harmonica and external α-helices. The active center, which includes cysteine from position 152 and histidine number 179, is responsible for the aforementioned NAD + binding. Rossmann folds, in addition to NAD + binding, are also responsible for RNA binding. However, this function is inactive during glycolysis. Such a mechanism of mutual inactivation enables the binding of the appropriate compound depending on its concentration in the environment. The catalytic (C-terminal) domain plays an important role in the course of the reaction. It is composed of 8 so-called β-sheets and α-helices attached to one side. Moreover, on the other side, the domain connects to its adjacent (more precisely, its α-helices). The task of this enzyme fragment is to bind G3P and to control the maintenance and localization of the protein inside the cell. The active site of the enzyme is located in a large gap between the catalytic domain and the coenzyme binding domain. This allows the simultaneous binding of G3P, NAD + and the phosphate ion (Muronetz et al. 2017).

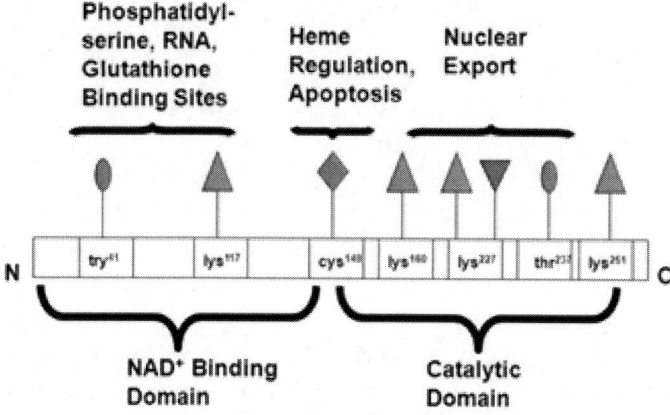

Figure 1. Scheme of the location of domains in GAPDH - N-terminal domain between the 1st and 150th amino acids and the C-terminal domain between the 151st and 335th amino acids. The figure shows the most important amino acids for the activity of the enzyme.

As mentioned in the introduction, GAPDH enables the conversion of G3P into 1,3-BPG in the glycolysis process (Figure 2). This is due to the sulfhydryl group of cysteine contained in the active center. The hemitioacetal resulting from the attachment of G3P to this group is then oxidized to the thiol ester. Histidine plays a special role in the first stage of this transformation, as it intensifies the above process. The previously mentioned ester undergoes further reactions (including phosphorylation), which leads to the production of 1,3-bisphosphoglycerate. The hydrogen obtained in the oxidation reaction, on the other hand, is attached to the NAD + located in the N-terminal domain. Completion of the process, assisted by His179, means the release of the transformation products and the reconstruction of the - SH group, which enables it to re-bind to glyceraldehyde-3-phosphate (Sirover et al. 2020).

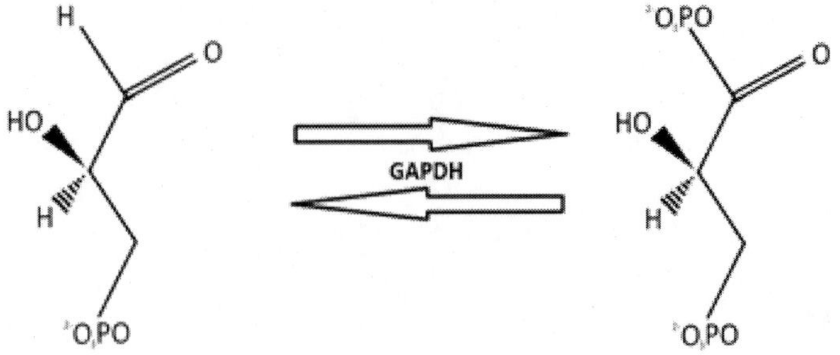

Figure 2. Scheme of the conversion of 3 phosphoglyceraldehyde into 1,3-bisphosphoglycerate.

Research on nerve cells (and then on hepatocytes and fibroblasts) at the beginning of the 21st century showed that GAPDH plays an auxiliary role in the process of apoptosis. It has been observed that this protein translocates to the cell nucleus and its accumulation in this organelle initiates apoptosis. Moreover, the chemical inactivation of the dehydrogenase causes the inhibition of programmed cell death. In the process of apoptosis with the participation of GAPDH, an important role is played by the process of acetylation and the addition of NO to the compound. Nitrosylation of dehydrogenase (or active site cysteine) allows it to bind to the siah-1 protein, and then transport the complex to the cell nucleus and induce apoptosis. This process is inhibited by the aforementioned compound - GOSPEL. It is a protein that binds to siah-1, thereby preventing it from binding to GAPDH and limiting the influx of dehydrogenase to the nucleus. Interestingly, the GOSPEL protein is most often found in muscles or the brain, i.e., in tissues with a high demand for energy. Acetylation mainly concerns the lysine residues (Lys 160) contained in the enzyme. The action of acetyltransferase induces the substitution of acetyl groups in nuclear proteins involved in apoptosis (e.g., p53, PUMA, p21, Bax). This process is a factor that initiates and regulates programmed cell death (White et al. 2017).

The protein described in the above chapter is a tetrameric enzyme that plays a special role in many processes, including glycolysis and apoptosis.

The specific structure of this enzyme (two domains) allows it to perform so many functions. During glycolysis, the N-terminal domain is active, which is involved in the attachment of NAD + to the active site, and the C-terminal domain (essentially the sulfhydryl group of cysteine) binding to glyceraldehyde 3-phosphate. Histidine also plays an important role in the subsequent stages of the glycolysis process.

Apoptosis is mainly based on acetylation and nitrolization. The simultaneous action of these two reactions allows not only the direct initiation of the described process, but also the stimulation of other accessory proteins.

In summary, G3P dehydrogenase is an enzyme that has long been known as a catalytic enzyme in the glycolysis process. It was not until the 20th century that detailed research into other functions of this protein began. The participation of GAPDH in apoptosis is one of the newest discoveries (around the first decade of the 21st century).

This enzyme is still not sufficiently understood and there are many questions that have not yet been answered. One of the as yet unexplored aspects (and clinically significant) is the use of GAPDH in cancer therapies.

The known features of dehydrogenase confirm its multifunctional character. It can be seen that it plays important roles almost throughout the life of the cell - from the formation of the cytoskeleton to its death.

References

Kosova AA, Khodyreva SN, Lavrik OI. Role of Glyceraldehyde-3-Phosphate Dehydrogenase (GAPDH) in DNA Repair. *Biochemistry* (Mosc). 2017 Jun;82(6):643-654.

Muronetz VI, Barinova KV, Stroylova YY, Semenyuk PI, Schmalhausen EV. Glyceraldehyde-3-phosphate dehydrogenase: Aggregation mechanisms and impact on amyloid neurodegenerative diseases. *Int. J. Biol. Macromol.* 2017 Jul;100:55-66.

Muronetz VI, Melnikova AK, Barinova KV, Schmalhausen EV. Inhibitors of Glyceraldehyde 3-Phosphate Dehydrogenase and Unexpected Effects of Its Reduced Activity. *Biochemistry* (Mosc). 2019 Nov;84(11):1268-1279.

Sirover MA. Moonlighting glyceraldehyde-3-phosphate dehydrogenase: posttranslational modification, protein and nucleic acid interactions in normal cells and in human pathology. *Crit. Rev. Biochem. Mol. Biol.* 2020 Aug;55(4):354-371.

White MR, Garcin ED. D-Glyceraldehyde-3-Phosphate Dehydrogenase Structure and Function. *Subcell Biochem.* 2017;83:413-453.

Chapter 6

Sucrose

Małgorzata Kraska, David Aebisher and Dorota Bartusik-Aebisher[*]
Medical College of the University of Rzeszów, Rzeszów, Poland

Abstract

Proteins play many important functions in the human body and amid the enormous amount of work they do, it is also worth looking at their enzymatic role in the reactions of decomposition of complex substances into simple substances.

Keywords: sucrose, glucose, digestive enzyme

Only simple substances, including glucose, can be used by humans in the process of obtaining energy. One such example is the saccharase discussed in the following section, which, as a digestive enzyme from the class of human disaccharidases, normally functions in the lumen of the small intestine in an optimal pH environment of about 4-5. Its name refers to the function it performs by catalysing the process of hydrolysis of sucrose to monosaccharides: glucose and fructose. Saccharase is part of the sucrase-isomaltase (SI) complex, where it is located as a subunit at the C-terminus and isomaltase on the other side (N-terminus). This complex belongs to the family 31 glycoside hydrolases (GH31). This family of enzymes is characterized by a series of proteins that are active against structures resulting from the

---

[*] Corresponding Author's Email: dbartusikaebisher@ur.edu.pl.

In: The Biochemical Guide to Enzymes
Editors: David Aebisher and Dorota Bartusik-Aebisher
ISBN: 979-8-88697-410-2
© 2023 Nova Science Publishers, Inc.

breakdown of starch. Several enzymes from GH31 are also present in the proteome of intestinal microbes (e.g., Bacteroides thetaiotamicron), indicating the complex and multi-faceted breakdown of saccharide compounds. SI deficiency is rare, occurs as a result of genetic mutations (e.g., autosomal recessive inheritance) and carries a number of serious symptoms, which will be detailed in detail discussed later in this chapter (Deb et al., 2021).

Saccharase is an enzyme discovered at the turn of the 19th and 20th centuries. Its discovery led to numerous experiments, among which we can distinguish groundbreaking, initial works:

- Kjeldahl's study took place in 1879 and 1881 and showed that sucrase has the ability to invert cane sugar, and this process is favored by the addition of small amounts of sulfuric acid.
- Fernbach's research took place in 1890, investigating the influence of various acids (especially organic ones) on the action of sucrase. It has been found that for each acid there is a certain range of optimal concentration within which a given enzyme exhibits maximum activity.
- Sörens' research took place in 1909 and focused on the study of the influence of hydrogen ions on the course of an enzymatic reaction.

The experiments of the above-mentioned scientists not only focused on studying the activity of the enzyme - saccharase, but also played a key role in providing information on the basics of enzymatic kinetics (Rathod et al., 2020).

It is a complex of enzymes (sucrase and isomaltase) associated with the membrane of the small intestine - it is a partially embedded integral protein located in the brush border. Only one subunit of the complex, isomaltase, interacts directly with the membrane via a highly hydrophobic segment in its N-terminal region. This segment crosses the membrane twice, with the N-terminus on the outside of the membrane. The sucrase subunit is attached to the membrane only through interaction with the isomaltase subunit. Due to the fact that the complex is tightly anchored to the brush border membrane, it has been classified as an integral type II membrane glycoprotein.

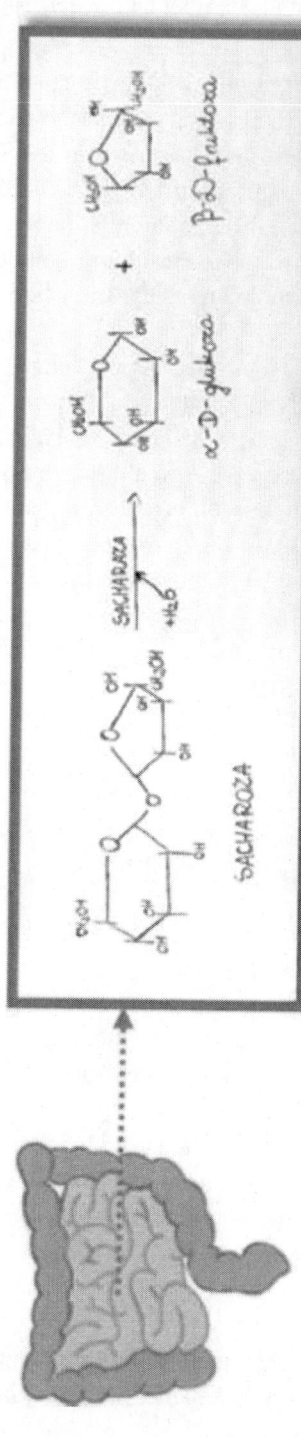

Figure 1. The mechanism of action of sucrase. Saccharase recognizes the α-1,2-glycosidic bond of sucrose, in which it catalyzes the hydrolysis to monosaccharides: glucose and fructose.

SI belongs to the classes of α-glucosidases, its role is the final breakdown of di- and oligosaccharides (hydrolysis of carbohydrates such as starch, saccharase, isomaltase) to digestible monosaccharides. This hydrolysis leads to the production of ATP after further processing. Of course, other enzymes of the GH31 family, e.g., glucoamylase, are also active in the lumen of the small intestine, but the role of SI is the most important because it is necessary for the final, key stage of carbohydrate digestion (e.g., dextrin hydrolysis) (Rose et al., 2018).

From amino acid sequence analysis, SI has been shown to arise from one GH31 precursor, which initially duplicates into the primary N-terminal and C-terminal domains, and then into two separate enzymes. As a result of heterodimerization, only the sucrase-isomaltase complex is formed, where sucrase is present at the C terminus, and isomerase - at the N terminus. Some studies have shown that sucrase may act as a chaperone chaperone for the pro-SI complex, and therefore may be the first intramolecular protein to locate at the C terminus instead of the standard N terminus. Importantly, both enzymes are enzymatically active, but the resulting pro-SI complex must be activated in the final SI by pancreatic proteases.

Saccharase is a unique enzyme due to its activity against the α-1,2-glycosidic disaccharide, which is sucrose (Figure 1). The structural basis of the ability of sucrase to hydrolyze the above-described linkage is to cleave the terminal glucose linked by an α-1,2-glycosidic bond to fructose. As a result of this transformation, α-D-glucose and β-D-fructose are formed (Kim et al., 2020).

Congenital sucrase-isomaltase deficiency (CSID) is a rare metabolic disorder of the gut with reduced or absent levels of sucrase-isomaltase (SI) activity. Interestingly, the main symptoms of CSID overlap with those of irritable bowel syndrome (IBS), a common functional gastrointestinal disorder of unknown etiology. The disease is inherited in an autosomal recessive manner and its prevalence is estimated at 0.2% of the Caucasian population. As a result of genetic mutations, the body lacks the enzyme saccharase necessary for the hydrolysis of sucrose in the small intestine. Clinical symptoms depend on the type of mutation, as heterozygous individuals usually do not suffer from disease manifestation. Sick people cannot consume not only sucrose and maltose, but also starch (SI is involved in digesting its breakdown products, i.e., dextrins). The first symptoms appear early - when the child begins to consume ready-made modified milk containing sucrose. After consumption, there are such inconveniences as: abdominal distension, excess gas or gushing diarrhea. Avoiding sucrose (present in, for example, fruit) allows you to free yourself from bothersome symptoms, and it has been found in children that with age the sugar tolerance improves (Tuck et al., 2019).

Summary

This chapter presents the most important aspects of the structure and function of an enzyme called sucrase, and shows the relationship between this theoretical information and its translation into clinical aspects that medical practitioners may encounter. The history of research leading to the discovery of sucrase and its biochemical properties is closely related to important discoveries for the newly emerging science of enzymology in the 19th century. Sucrase is properly produced in the brush border of the small intestine, where, at the appropriate pH, it has the ability to break down disaccharides (such as sucrose) into fructose and glucose, which can be converted into carbon dioxide, water and energy in the form of ATP in the further process of intracellular respiration.

An important issue discussed in the chapter was the relationship between the presence of sucrase in a complex with isomaltase, forming an SI complex, where sucrase locates at the C terminus and acts as a chaperone (chaperone) - this is interesting as chaperone proteins usually locate at the N terminus; moreover, both subunits are structurally similar, and these two issues are the subject of much contemporary research. SI deficiency, called CSID, is the key to diagnosis in the early years of the patient's life by a clinician and the introduction of an appropriate diet low in complex sugars such as sucrose, maltose and starch for a given person. By following a diet low in these complex carbohydrates, symptoms can be relieved (symptomatic treatment), but genetic research is underway to help with complete recovery.

References

Deb C, Campion S, Derrick V, Ruiz V, Abomoelak B, Avdella A, Zou B, Horvath K, Mehta D I. Sucrase-isomaltase Gene Variants in Patients With Abnormal Sucrase Activity and Functional Gastrointestinal Disorders. *J. Pediatr. Gastroenterol. Nutr.* 2021 Jan 1;72(1):29-35.

Kim S B, Calmet F H, Garrido J, Garcia-Buitrago M T, Moshiree B. Sucrase-Isomaltase Deficiency as a Potential Masquerader in Irritable Bowel Syndrome. *Dig. Dis. Sci.* 2020 Feb;65(2):534-540.

Rathod S, Friesen C A, Radford K, Colombo J M. Sucrase Breath Testing in Children Presenting With Chronic Abdominal Pain. *Clin. Pediatr.* (Phila). 2020 Nov;59(13): 1191-1194.

Rose D R, Chaudet M M, Jones K. Structural Studies of the Intestinal α-Glucosidases, Maltase-glucoamylase and Sucrase-isomaltase. *J. Pediatr. Gastroenterol. Nutr.* 2018 Jun;66 Suppl 3:S11-S13.

Tuck C J, Biesiekierski J R, Schmid-Grendelmeier P, Pohl D. Food Intolerances. *Nutrients.* 2019 Jul 22;11(7):1684.

Chapter 7

Dehydrogenases

Julia Motyka, David Aebisher and Dorota Bartusik-Aebisher[*]
Medical College of the University of Rzeszów, Rzeszów, Poland

Abstract

Dehydrogenases are a group of proteins that act as a catalyst, which means that by influencing the reaction, they do not change the composition of the final mixture or the equilibrium constant, but only accelerate the equilibrium state of the system by lowering the activation energy value.

Keywords: dehydrogenases, Krebs cycle, biochemistry

Dehydrogenases belong to the first class of enzymes - oxidoreductases. They are characterized by a different structure or properties, but they have the same function - they catalyze the hydrogen atom decoupling reaction - the reduction or oxidation reaction (such as, for example, alcohol dehydrogenase, which can carry out both the oxidation reaction of the alcohol to an aldehyde or ketone and the reduction of the aldehyde to alcohol). These reactions require the use of coenzymes (NAD^+, $NADP^+$ or flavin coenzyme) to which protons and electrons are transferred (Adachi et al. 2019).

Dehydrogenases are involved in important processes, including the metabolism of fatty acids, and are also catalysts for many stages of the Krebs cycle, such as isocitrate dehydrogenase, α-ketoglutarate dehydrogenase,

---

[*] Corresponding Author's Email: dbartusikaebisher@ur.edu.pl.

In: The Biochemical Guide to Enzymes
Editors: David Aebisher and Dorota Bartusik-Aebisher
ISBN: 979-8-88697-410-2
© 2023 Nova Science Publishers, Inc.

succinate dehydrogenase or malate dehydrogenase, while glucose-6-phosphate dehydrogenase protects cells against oxidative stress. They are also of clinical importance, an example is lactate dehydrogenase, the excessive amount of which may indicate an early stage of myocardial infarction or increased hemolysis, and aldehyde or alcohol dehydrogenase, and an increase or decrease in their concentration is important in laboratory diagnostics - some of them may be secreted by cancer cells or increase their levels in other disease states (Adachi et al. 2019).

Dehydrogenases, although they belong to one group of enzymes, differ in structure and properties, such as the Michaelis constant or the optimal pH range, which results from different sites of enzyme activity. LDH can be found in red blood cells or myocardium, where it is secreted in damage, and is an indicator of tissue damage, such as, for example, myocardial infarction, while the site of ADH action is mainly in the liver, where alcohol is metabolized, as schematically shown in Figure 1, but also in the brain, where it is involved in the metabolism of substances such as retinol or serotonin and in other organs. Differences in the optimal pH range may also occur within one enzyme, depending on whether the oxidation or reduction reaction is catalyzed (Csarman et al. 2020).

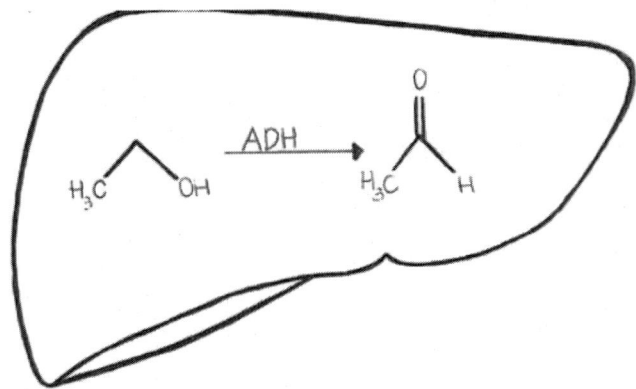

Figure 1. Schematic representation of the metabolism of ethyl alcohol. Ethanol is converted to acetaldehyde under the influence of alcohol dehydrogenase; the majority of the process takes place in the liver [own elaboration].

The diversity of dehydrogenases consists not only in their different types associated with a broad substrate spectrum, but also in many variants of one enzyme: isoenzymes, encoded by various alleles and characterized by different degrees of activity. Examples include the genes encoding the isoenzymes of alcohol and aldehyde dehydrogenase. ADH1B and ADH1C are responsible for the highly active form of ADH, while $ALDH_2$ is responsible for the inactive form of ALDH. The presence of both of these forms results in the accumulation of acetaldehyde, which may be of clinical significance due to the reduction of the risk of developing ADH. alcoholism. Dehydrogenases are characterized by stereospecificity; this means that the hydrogen atom from the substrate can be transferred either to the pro-R position (where one hydrogen atom connected to the fourth carbon atom is directed to the substrate binding site, and the other to the protein), or to the pro-R position. S (it is the opposite conformation to pro-R dehydrogenase) of the fourth carbon atom of the nicotinamide ring (Madak et al. 2019).

Oxidation of the substrate by dehydrogenases occurs by transferring a hydrogen atom to an electron acceptor, which are the aforementioned coenzymes, and then a water molecule associated with the enzyme can be attached. The result of such a reaction can be a positive charge for the substrate or the formation of a double bond. For the formation of a double bond, two hydrogen atoms must be cleaved off, and the role of a cofactor is played by FAD.

The reduction reaction consists in transferring the hydrogen atom of the reduced coenzyme to the substrate - in the case of ADH to the C = O bond. The oxidases are characterized by a similar principle of operation, but they differ in their electron acceptor. In the case of oxidases it can only be molecular oxygen (Porzani et al. 2021).

The type of coenzyme - also known as an electron acceptor - involved in the reaction depends on the reaction site and the substrates. NAD^+, nicotinamide adenine dinucleotide, consists of two nucleotides consisting of the adenine group and nicotinamide. During reduction, a proton and two electrons are added to the six-carbon nicotinamide ring. It is possible to determine the activity of a given dehydrogenase by spectrophotometric analysis of the NAD^+ coenzyme, because its reduced form shows an additional absorption band at a wavelength of 340 nm.

Another coenzyme is $NADP^+$, which additionally has a phosphate group that allows it to bind to the active center of another enzyme. Some enzymes are able to catalyze the reaction in both the presence of NAD^+ and $NADP^+$, such as for example glutamate dehydrogenase.

FAD, or flavin dinucleotide, is a prosthetic group, so unlike the coenzyme, it is permanently bound to the enzyme, ensuring its proper functioning. It consists of adenine and flavin mononucleotide. It is joined by two hydrogen atoms, so its reduced form is $FADH_2$ (Scheiblbrandner et al. 2020).

Dehydrogenases are a group of enzymes with a broad substrate spectrum that are active in many tissues and organs of the body. They are an indicator of cell activity and may indicate pathological processes occurring in tissues, which affects their diagnostic significance. They also exist in the form of isoenzymes with different properties. They also differ in the spatial arrangement - they can occur in pro-S or pro-R conformations, resulting from their location in relation to the coenzyme. They are also characterized by different reaction conditions, and thus a different value of the optimal pH range, which may differ in the case of oxidation and reduction reactions catalyzed by one type of dehydrogenase.

These enzymes can catalyze the reduction and oxidation reactions by transferring the hydrogen atom to the appropriate cofactors acting as proton and electron acceptors, which may be NAD +, NADP + and FAD, while the presence of an oxidized or reduced form of the coenzyme enables the determination of dehydrogenase activity using a spectrophotometer. The vast majority of dehydrogenases are capable of interacting with only one cofactor, but some have the ability to use several types of them. These reactions involve the cleavage of the hydrogen atom from the substrate and can take place with the acquisition of a positive charge or with the formation of a double bond.

References

Adachi T, Kaida Y, Kitazumi Y, Shirai O, Kano K. Bioelectrocatalytic performance of d-fructose dehydrogenase. *Bioelectrochemistry*. 2019;129:1-9.

Csarman F, Wohlschlager L, Ludwig R. Cellobiose dehydrogenase. *Enzymes*. 2020;47:457 489.

Madak JT, Bankhead A 3rd, Cuthbertson CR, Showalter HD, Neamati N. Revisiting the role of dihydroorotate dehydrogenase as a therapeutic target for cancer. *Pharmacol. Ther.* 2019;195:111-131.

Porzani SJ, Lorenzi AS, Eghtedari M, Nowruzi B. Interaction of Dehydrogenase Enzymes with Nanoparticles in Industrial and Medical Applications, and the Associated Challenges: A Mini-review. *Mini Rev. Med. Chem.* 2021;21(11):1351-1366.

Scheiblbrandner S, Ludwig R. Cellobiose dehydrogenase: Bioelectrochemical insights and applications. *Bioelectrochemistry*. 2020;131:107345.

Chapter 8

Telomerase

Katarzyna Koszarska, David Aebisher and Dorota Bartusik-Aebisher*
Medical College of the University of Rzeszów, Rzeszów, Poland

Abstract

Telomerase is a large enzyme with the structure of a ribonucleoprotein complex (RNP), which is responsible for the aging of cells, and thus the entire organism. The function of this enzyme is similar to RNA-dependent DNA polymerase.

Keywords: telomerase, enzyme, biochemistry

The site of action of telomerase are telomeres - the end sections of chromosomes, which in humans constitute up to several thousand repeats (5-15 kb) of the 5 '- TTAGGG - 3' sequence. Both strands of DNA are shortened in most cells successively with each cell division by about 50 nucleotides, in connection with the so-called end of replication problem. This problem is due to the lack of enzymatic functions in most human eukaryotic cells to add missing nucleotide sequences on the delay strand, after removing the last primer at the 3 'end of this strand. Therefore, on the leading strand, which divides, the DNA section is extended without complementary reflection on the delayed strand - it is missing in these places. Such lengths on the leading thread are referred to as overhangs (Colella et al. 2019). Then these free fragments

---

* Corresponding Author's Email: dbartusikaebisher@ur.edu.pl.

In: The Biochemical Guide to Enzymes
Editors: David Aebisher and Dorota Bartusik-Aebisher
ISBN: 979-8-88697-410-2
© 2023 Nova Science Publishers, Inc.

fold, adjusting their sequence to the sequences earlier within dsDNA and join at this point for a length of several or a dozen nucleotides (the second strand of the helix at this short junction point remains disconnected, forming the so-called D-loop (smaller loop). but this is not a conflict in gene expression, since these sites do not contain valuable genetic information anyway - their function is to protect the middle parts of the chromosomes). Such a curl creates a characteristic cap, the so-called T-loop (larger loop) securing telomeres against damage or the effects of mutagenic factors (Meena et al. 2020).

Telomerase is an enzyme found in eukaryotic cells, but in most of them its function is lost due to the action of the repressor, it occurs passively as an inactive enzyme, so the cell division capacity is limited to approx. 50-70 divisions. It can be found physiologically active in:

- Certain stem cells
- Germ line cells

While pathologically it occurs in 85% of neoplastic cells and is one of the reasons for their uncontrolled division. As a result, telomerase is a very promising object of clinical research on the proliferative activity of these cells. The aim of the research is to find an inhibitor of active telomerase. An important aspect in this regard is the fact that the telomeres of neoplastic cells are much shorter than the chromosomal telomeres contained in the cells of the human body. This means that the intensity of inhibitors of this enzyme causing complete shortening of cancer cells' telomeres and, as a result, their apoptosis (when telomeres are critically shortened, they lose their ability to produce a T loop, which signals the cell to go into the apoptotic pathway), will not have an apoptotic effect on human cells, germline cells or stem cells of considerable importance. The loss of their properties in favor of the elimination of neoplastic cells would not be compensatory. Potential inhibitors are low molecular weight compounds, structurally analogous to nucleotides and nucleosides. In another aspect, if a mechanism stimulating the activity of this enzyme in somatic cells was developed, it would be possible to correct degenerative pathological changes, or even to control the aging of tissues. Finding and carefully developing the appropriate mechanisms would be the basis of scientific fields such as oncology and aesthetic medicine, not to mention the cosmetics industry. So far, research shows that the most promising activators may be (similarly to inhibitors) low molecular weight compounds - e.g., a saponin-like substance isolated from the root of Astragalus membranaceus - membranous astragalus (Mehraban et al. 2019).

Structurally, human telomerase is a complex enzyme complex with many important components, the most important of which are:

- TERT subunit, a catalytic region with the function of reverse transcriptase;
- TER subunit, RNA acting as a template;
- Dyskerin and many other proteins responsible for the stabilization of the structure.

Telomerase is classified as a dimer and its length is estimated to be around 28nm. In human cells, it occurs in the so-called Cajal bodies (CB) - where their synthesis also takes place - are granules 0.1 - 0.2 µm in size, contained in the nucleus of proliferating cells. The number of Cajal bodies is a variable value and increases in the G1 phase of the cycle, while there are still unconfirmed hypotheses that the complete disappearance of the enzyme is observed in the S phase (Visnapuu et al. 2019).

The enzyme cycle of action includes several steps: connection to the primer site, elongation, translocation and dissociation. The RNA sequence contained in the enzyme matches the final nucleotides to the overhang sequence (on the leader strand) and the remaining free nucleotides cause the strand to be elongated by the replication mechanism (the RNA fragment contained is a template), which then enables the insertion of the primer into the elongated leading strand through the primase and the addition of the second strand - resynthesis of the lost dsDNA structure at the ends of telomeres by the action of α-polymerase.

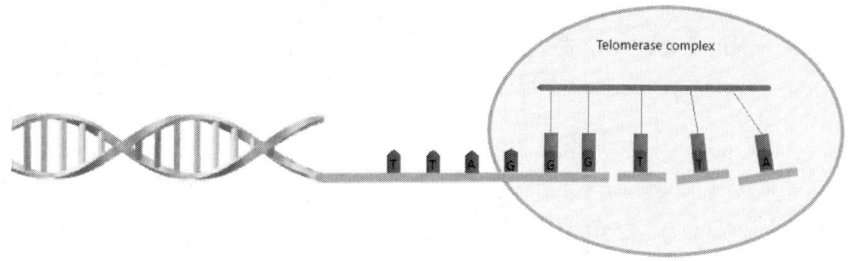

Figure 1. Function of the telomerase complex.

In embryonic life, the telomerase present in cells shows an increased activity, which decreases shortly after birth, until it is completely repressed in most cells. The new organism has the so-called reset telomere clock. Diseases associated with disturbed telomerase activity in utero are, for example, congenital dyskeratosis or aplastic anemia. It may also be associated with genome-wide oncogenic instability and telomeropathies in the postnatal life. Unfortunately, research on telomerase, despite its importance, remains very difficult due to the low content of the enzyme in cells and problems with its isolation (Zhang et al. 2019).

A great deal of research on telomerase has been done in yeast because it is easier to isolate from cells. Work on the enzyme in yeast makes it possible to approximate, for example, the process of joining the ribonucleoprotein complex to telomers in DNA strands, where it was found that assembly of complete RNP takes place only after replication in these organisms. In humans, previous studies have ruled out similarities in these aspects, while many similarities in yeast may significantly translate into the functioning of this enzyme in humans. The problem in studying the work of this enzyme is the fact that various substances, still little known, are responsible for various stages of its action, e.g., a given molecule in vitro recruits the enzyme at a specific locus on the telomere, but no longer has the ability to activate it in vivo.

The very efficiency of the DNA strand addition process is very variable in telomerase-active cells and depends on many factors, such as: the number of enzyme molecules in the cell, the degree of interaction of these molecules with the specific site of action and the sum of the repeats added by the enzyme before dissociation (which depends on the main measure on the speed of its operation). The sudden increase in interest in telomerase in recent years has made people realize the clinical importance of this complex. The scope of conducted research is expanding dynamically, the development of modern methods such as super resolution microscopy gives hope for a much more detailed understanding of this enzyme, which is of great importance for the properties of cancer cells. Certainly, there will come a time when research on it develops to such an extent that it will be possible to identify its full mechanism of action and establish an effective inhibition and activation mechanism, usable on a large scale, which would be a breakthrough in anti-cancer treatment and more. Stem cell research remains promising here.

References

Colella P, Mingozzi F. Gene Therapy for Pompe Disease: The Time is now. *Hum Gene Ther.* 2019;30(10):1245-1262.

Meena NK, Raben N. Pompe Disease: New Developments in an Old Lysosomal Storage Disorder. *Biomolecules.* 2020;10(9):1339.

Mehraban MH, Mansourian M, Ahrari S, HajiEbrahimi A, Odooli S, Motovali-Bashi M, Yousefi R, Ghasemi Y. Maltase-glucoamylase inhibition potency and cytotoxicity of pyrimidine-fused compounds: An in silico and in vitro approach. *Comput Biol Chem.* 2019;82:25-36.

Visnapuu T, Meldre A, Põšnograjeva K, Viigand K, Ernits K, Alamäe T. Characterization of a Maltase from an Early-Diverged Non-Conventional Yeast *Blastobotrys adeninivorans*. *Int J Mol Sci.* 2019;21(1):297.

Zhang M, Wang H, Wang B, Ma Y, Huang H, Liu Y, Shao M, Yao B, Kang Z. Maltase Decorated by Chiral Carbon Dots with Inhibited Enzyme Activity for Glucose Level Control. *Small.* 2019;15(48):e1901512.

Chapter 9

Oxoglutarate Dehydrogenase

Karolina Lach, David Aebisher and Dorota Bartusik-Aebisher[*]

Medical College of the University of Rzeszów, Rzeszów, Poland

Abstract

Proteins are essential for the proper functioning of the cell and the entire organism. Enzymes responsible for the fast and coordinated course of metabolic and biochemical changes, crucial for meeting the energy needs of cells, are of particular importance. Such proteins include oxaglutarate dehydrogenase.

Keywords: 2-Oxoglutarate dehydrogenase, Krebs cycle, enzymes

2-Oxoglutarate dehydrogenase (OGDH), otherwise known as α-ketoglutarate dehydrogenase, plays a significant role in cellular metabolism. It is part of the 2-oxoglutarate dehydrogenase complex (OGDHC), which is an important element of the citric acid (Krebs) cycle. The other components of the OGDHC are succinyl dihydroliponate transferase and dihydroliponate dehydrogenase. Together with them, in the form of repetitive copies, it forms the spherical structure of the OGDHC. This complex catalyzes the reaction of 2-oxoglutarate to succinyl-CoA. In addition, it takes part in the pathways of transformation of amino acids, fatty acids and carbohydrates (Choi et al. 2019).

---

[*] Corresponding Author's Email: dbartusikaebisher@ur.edu.pl.

In: The Biochemical Guide to Enzymes
Editors: David Aebisher and Dorota Bartusik-Aebisher
ISBN: 979-8-88697-410-2
© 2023 Nova Science Publishers, Inc.

The first mentions of OGDH in the scientific literature can be noticed already in the 1950s, when research on the isolation of α-ketoglutarate dehydrogenase was carried out on pig heart muscle cells. There is currently an interest in oxaglutarate dehydrogenase in detecting the association of the above protein with cancer development, neurodegenerative diseases such as Alzheimer's disease, and neurological disorders. One of the most recent papers on oxoglutarate dehydrogenase published this year addressed the clinically important issue of the effect of OGDH on sepsis-induced acute lung injury (ALI). This study could have a significant impact on potential treatment options for ALI patients, as it is a common cause of death for which effective treatments are still lacking (Jordan et al. 2019).

Despite the fact that 2-oxoglutarate dehydrogenase is a protein known for many years, numerous studies are still carried out today and new relationships are discovered related to its activity and the development of many disorders of homeostasis and diseases of the organism. Over the past three years, new studies have been made, three of which have been selected, and the key aspects of these studies are presented in the following article.

A common mechanism that can trigger ALI in animals is the ligation and puncture of the cecum (CLP). It was used by a group of researchers and the following results of their work were published this year. Studies have shown that there is an increased transcription of oxoglutarate dehydrogenase in CLP-induced ALI. The OGDHC complex is an enzyme active mainly in the mitochondria. It catalyses the conversion of 2-oxoglutarate into succinyl-CoA and CO_2. Its malfunction leads to a disturbance of the mitochondrial function, as a result of which the amount of reactive oxygen species increases and cells apoptosis occurs. This may be one of the causes of organ damage in sepsis. Multiple organ failure may develop as a result of sepsis. The lungs are particularly vulnerable organs due to their delicate structure and slow blood flow in their vessels, which results in longer exposure to toxic substances and inflammatory mediators. As a result of sepsis, many patients develop ALI or acute respiratory distress syndrome. On the basis of numerous tests and research procedures, it has been established that OGDH induces lung damage due to an increase in cytokine concentration in the inflammatory process and is regulated by the MAPK signal transduction pathway. The above study offers hope for the discovery of a potentially effective therapeutic method in the course of acute pneumonia (Losman et al. 2020).

Even a small mutation in the genetic material can cause significant changes in the functioning of the organism. In many cases, a significant defect is lethal already during embryogenesis and fetal life. However, if the pregnancy does not end in a miscarriage or stillbirth, the genetic disorder will accompany the baby in the next life, often showing up unexpectedly during development. In 2021, a publication was published by a group of researchers working on discovering the cause of neurological disorders that occurred in two out of five siblings, in a family of Syrian origin, whose parents were related to each other. In the case of an older patient, up to 8 months of age, when the child fell down without loss of consciousness or intracranial bleeding, no developmental abnormalities were observed. From that moment on, she lost her ability to sit and crawl, failed to learn to speak, developed dystonic movements, and was moving in a wheelchair (Morris et al. 2019).

In the case of the younger patient, the first symptoms indicating the development of the disease were movement disorders, manifested by inability to run and numerous falls, as well as motor disorders and speech delay. Most of the commissioned studies did not indicate any major irregularities. Both patients had alternately high levels of lactate. On the basis of numerous genetic studies, it has been concluded that it is possible that this is the first case of a pathogenic variant (c.959A> G) in the OGDH gene, which resulted in a pathogenic amino acid substitution (p.N320S). This mutation likely led to the development of the above disease. The other three siblings showed no disturbing symptoms (Losman et al. 2020).

Breast cancer is characterized by a high incidence and about 20% frequency of metastases, mainly to the lungs, bones and soft tissues. Despite the many preventive measures and existing therapies, there is still a need for treatments to prevent metastasis. Cancer cells require intensive metabolic processes, including in the mitochondria. As mentioned before, the oxoglutarate dehydrogenase complex (OGDHC) is an important mitochondrial enzyme. A study published in 2018 on the effect of inhibition of α-ketoglutarate dehydrogenase by the action of (S) -2 - [(2,6-dichlorobenzoyl) amino] succinic acid (AA6) shows that it is a promising drug, capable of inhibiting further development breast cancer that has spread to other tissues. The above relationship was confirmed experimentally by examining two human breast cancer cell lines (CRL-2335 and MDA-231) and on 4T1 cells, where KGDH was reduced by the CRISPR / Cas9 method. Moreover, the results indicated an interaction between metabolic changes taking place in the mitochondria and the mechanisms regulating the process of nuclear transcription (Xu et al. 2019).

This article presents some of the recent studies on 2-oxoglutarate dehydrogenase. It is an enzyme essential in the proper course of the Krebs cycle, and is a source of reactive oxygen species (ROS) in the mitochondria.

However, based on the course of studies published last year, a certain exception was discovered in the cells of the heart muscle. It was found that the dehydrogenase α-Ketoglutarate is not the main source of ROS in this case and is even possible that it contributes to the elimination of H_2O_2. The 2-oxoglutarate dehydrogenase complex is considered to be one of the major regulators of mitochondrial metabolism due to its association with the amount of NADH and reactive oxygen species found in these cell organelles.

A relationship has also been demonstrated between decreased OGDHC activity, associated with impaired mitochondrial function, and the development of neurodegenerative diseases. OGDHC is a macromolecular complex, particularly active in ensuring the proper level of regulation of metabolic pathways and the catalytic efficiency of the cell. Moreover, the potentially important role of OGDH in the treatment of neoplastic diseases cannot be overlooked. It has been proven on the basis of numerous studies that there are connections between the proper functioning of oxaglutarate dehydrogenase, especially its participation in the course of the Krebs cycle, and the homeostasis of various types of neoplastic cells.

Figure 1. 2 - oxoglutarate dehydrogenase (α - ketoglutarate dehydrogenase) in the Krebs cycle.

Due to the high interest in OGDH activity that has emerged in recent years, there is hope for the emergence of new effective methods of treating diseases such as Alzheimer's disease, some of the neurological disorders and neoplastic diseases, ALI and many others, the etiology of which is related to the action of oxoglutarate dehydrogenase.

References

Choi S, Pfleger J, Jeon YH, Yang Z, He M, Shin H, Sayed D, Astrof S, Abdellatif M. Oxoglutarate dehydrogenase and acetyl-CoA acyltransferase 2 selectively associate with H2A.Z-occupied promoters and are required for histone modifications. *Biochim Biophys Acta Gene Regul Mech.* 2019;1862(10):194436.

Jordan F, Nemeria N, Gerfen G. Human 2-Oxoglutarate Dehydrogenase and 2-Oxoadipate Dehydrogenase Both Generate Superoxide/H_2O_2 in a Side Reaction and Each Could Contribute to Oxidative Stress in Mitochondria. *Neurochem Res.* 2019;44(10):2325-2335.

Losman JA, Koivunen P, Kaelin WG Jr. 2-Oxoglutarate-dependent dioxygenases in cancer. *Nat Rev Cancer.* 2020;20(12):710-726.

Morris JP 4th, Yashinskie JJ, Koche R, Chandwani R, Tian S, Chen CC, Baslan T, Marinkovic ZS, Sánchez-Rivera FJ, Leach SD, Carmona-Fontaine C, Thompson CB, Finley LWS, Lowe SW. α-Ketoglutarate links p53 to cell fate during tumour suppression. *Nature.* 2019;573(7775):595-599.

Xu Y, Shen J, Ran Z. Emerging views of mitophagy in immunity and autoimmune diseases. *Autophagy.* 2020;16(1):3-17.

Chapter 10

Maltase

Aleksandra Kotlińska, David Aebisher and Dorota Bartusik-Aebisher[*]
Medical College of the University of Rzeszów, Poland

Abstract

Without the presence of numerous enzymes, digesting carbohydrates, one of the main components of our diet, would not be possible. Between monosaccharide molecules there are usually α-glycosidic bonds, and less frequently, as in the case of lactose β-glycosides, which enable the formation of complex carbohydrates, taking the form of long, sometimes branched chains.

Keywords: β-glycosides, hydrolysis of α-glycosidic, sucrase-isomaltase

Hydrolysis of α-glycosidic bonds begins in the oral cavity under the influence of salivary amylase and continues in the small intestine due to the presence of pancreatic amylase. In this way, disaccharides are formed from complex molecules, which are then treated with the glycosidases present within the brush border, such as maltase-glucoamylase (MGAM) and sucrase-isomaltase (SI) (Goda et al. 2018). In this way, complex sugars are transformed into simple sugars that can be absorbed into enterocytes by counter-transporting with Na^+ ions (SGLT1 cotransporter) or via GLUT5 transporters (Goda et al. 2018).

---

[*] Corresponding Author's Email: dbartusikaebisher@ur.edu.pl.

In: The Biochemical Guide to Enzymes
Editors: David Aebisher and Dorota Bartusik-Aebisher
ISBN: 979-8-88697-410-2
© 2023 Nova Science Publishers, Inc.

Maltase is an enzyme belonging to hydrolases, it consists of 1,857 amino acids, and the gene that codes for it is located on chromosome 7 and contains over 82,000 base pairs. Two domains can be distinguished in it: the N-terminal with high activity against maltose, and the domain C-terminal with a broader substrate spectrum, active against glucose oligomers.

This enzyme is one of the key elements of the carbohydrate digestion pathway, the disturbance in the expression or abnormal structure of this protein causes the malabsorption of carbohydrates, which has many consequences. The sugars remaining in the digestive tract are too large to be absorbed, so there is an increase in osmotic pressure in the gastrointestinal tract, which results in an inflow of water and the resulting diarrhea (Lee et al. 2018).

Thus, in summary, understanding the proper structure and function of this enzyme is essential to understanding the causes of nutrient malabsorption diseases, which will enable the most effective therapy for the patient to be tailored.

The maltase-glucoamylase consists of five distinct domains (shown in Figure 1):

- N-terminal domain
- C-terminal domain
- Small cytoplasmic domain (approx. 26 amino acids)
- Transmembrane domain (approx. 20 amino acids)
- O-gicosidated linker domain (approx. 55 amino acids)

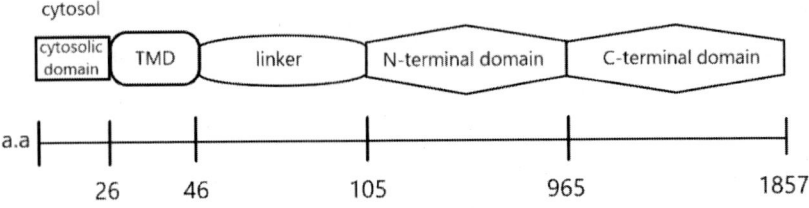

Figure 1. Schematic diagram of the MGAM structure.

The most important for the enzyme activity are the N-terminal (MGAM-N) and C-terminal (MGAM-C) domains, which determine the ability to hydrolyze linear 1,4-glycosidic bonds, therefore they will be described in more detail. These subunits are homologous and due to their activity are classified as glycoside hydrolases, they catalyze the same type of reaction, but differ in substrate specificity (Meena et al. 2020).

MGAM-N has the greatest activity against substrates containing two glucose molecules, contains 21 amino acids less than the MGAM-C subunit.

MGAM-C is active against oligosaccharides with up to four glucose molecules, consists of five parts: P-type (cysteine-rich) domain, N-terminal domain, catalytic domain, proximal C-terminal domain and distal C-terminal domain.

As mentioned earlier, maltase hydrolyzes linear bonds, however, as a result of a single point mutation in the gene encoding the MGAM-C subunit (Trp1251), the enzyme may gain the ability to break down 1,6-glycosidic branched bonds (Meena et al. 2020).

Due to the type of catalyzed reaction, inhibition of maltase may be used in the treatment of diseases such as obesity and type II diabetes. Treatment of type II diabetes mellitus is aimed at preventing the blood glucose level from rising too much after eating (postprandial hyperglycemia). Acarbose and miglitol are drugs that inhibit α-amylase present in pancreatic juice and α-glycosidases bound in the enterocyte membrane (Nichols et al. 2018). These preparations slow down and reduce the amount of broken-down carbohydrates, hence more of them enter the large intestine and then excreted with the faeces. The listed drugs have a beneficial effect in the treatment of diabetes, but they cause unpleasant side effects, such as gastrointestinal ailments (Nichols et al. 2018).

Natural glycosidase inhibitors have been found in Salacia reticulata - a plant found in Sri Lanka, India, these are salacinol and kotalanol. Research on synthetic analogues of these compounds is ongoing, and a lot of attention is paid to blintol, an analog of salicinol, which in a study on rats shows the effect of regulating blood glucose after a meal rich in carbohydrates, which makes it useful in the treatment of type II diabetes.

The maltase-glucoamylase deficiency was developed from studies in mice. They showed that with the damaged membrane part of MGAM, the activity of maltase decreased by 40%, with a simultaneous, compensatory increase in the activity of SI, which ensures the digestion of sugars.

In another study, it was noticed that when feeding mice with C^{13}-labeled starch in subjects with reduced MGAM activity, the concentration of labeled glucose in the blood was lower than in subjects with normal activity of this enzyme, but no differences were found between the total glucose concentration in mice from the test sample (damaged MGAM) and a control sample (correct MGAM). The same study noted that the concentration of insulin in the blood of mice with damaged MGAM is lower, which indicates that MGAM has an effect on the insulin response to starch-rich food (Rose et al. 2018).

The visible paradox between exogenous starch glucogenesis, and total blood glucose turnover suggested that MGAM may have an effect on mealtime control of endogenous gluconeogenesis. It has been hypothesized that MGAM is more important for glucogenesis with limited access to starch-rich food, while with unlimited access to food, SI activity dominates. This was evidenced by a 40% reduction in blood glucose only in mice with reduced MGAM activity and limited access to food, while with unrestricted access to food, the glucose level was the same in both mice with normal MGAM activity and mice with reduced activity of this enzyme.

In humans, the activity of this enzyme is tested by biopsy, i.e., taking a small (<10 mg) fragment of the duodenal mucosa. Deficiency is defined by too low enzyme activity, below the lower limit of the norm, which for maltase is 100 μmol^{-1} min^{-1} g^{-1} (Rose et al. 2018).

Maltase-glucoamylase is an enzyme that determines the proper absorption of nutrients, as in the case of other compounds of this type, it has a specific active site thanks to which it can bind the appropriate substrate, in the case of maltase the bound molecules are disaccharides, which are then broken down into simple sugars. In this way, maltase is included in the food digestion pathway, the proper course of which is essential to obtain the energy needed for every cell in our body.

Due to the characteristic structure of enzymes, their inhibition may occur, and this is also the case with maltase. Compounds that inhibit its effect are, for example, acarbose and miglitol, as well as salacinol and kotalanol, which are still being researched. Cases of decreased activity or deficiency of this enzyme are rarely recognized, as it has been proven in studies, the reduction in maltase activity may be compensated by sucrase-isomaltase, which is also classified as glycosidase, in addition, the symptoms of such a deficiency are not clear, as diarrhea, abdominal distension or indigestion may indicate many other diseases. The inhibition of maltase may also bring beneficial effects, especially in the treatment of diabetes, the reduced activity of the enzyme causes that after a meal, less disaccharides is broken down into simple sugars, and thus less glucose goes to the blood, which is the main goal of therapy in type 2 diabetes.

Knowing how maltase and other enzymes work allows you to do it faster and the correct diagnosis of diseases resulting from the abnormal functioning of this protein, as well as enabling research on the use of this compound in the treatment of other diseases.

References

Goda T, Honma K. Molecular Regulations of Mucosal Maltase Expressions. *J Pediatr Gastroenterol Nutr.* 2018 Jun;66 Suppl 3:S14-S17.
Lee BH, Hamaker BR. Maltase Has Most Versatile α-Hydrolytic Activity Among the Mucosal α-Glucosidases of the Small Intestine. *J Pediatr Gastroenterol Nutr.* 2018 Jun;66 Suppl 3:S7-S10.
Meena NK, Raben N. Pompe Disease: New Developments in an Old Lysosomal Storage Disorder. *Biomolecules.* 2020 Sep 18;10(9):1339.
Nichols BL, Baker SS, Quezada-Calvillo R. Metabolic Impacts of Maltase Deficiencies. *J Pediatr Gastroenterol Nutr.* 2018 Jun;66 Suppl 3:S24-S29.
Rose DR, Chaudet MM, Jones K. Structural Studies of the Intestinal α-Glucosidases, Maltase-glucoamylase and Sucrase-isomaltase. *J Pediatr Gastroenterol Nutr.* 2018 Jun;66 Suppl 3:S11-S13.

Chapter 11

Aminolevulinic Acid Synthase

Anita Majda, David Aebisher and Dorota Bartusik-Aebisher[*]
Medical College of the University of Rzeszów, Rzeszów, Poland

Abstract

ALA synthase, or aminolevulinic acid synthase (ALAS), is an enzyme that occurs naturally in the cells of our body, it is precisely present in the mitochondria. This protein is involved in the synthesis of porphyrin, which is part of the heme. Almost all living organisms need heme, which together with globin forms hemoglobin - an oxygen transporter.

Keywords: aminolevulinic acid synthase, enzyme, biochemistry

The biosynthesis of this compound must be tightly regulated in all cells, because both the lack and the excess of it cause many diseases in humans. During erythropoiesis, heme synthesis is intensified, and thus the activity of ALA synthase involved in this process is increased. Thus, in the case of a deficiency of aminolevulinic acid synthase, we can observe a decrease in the amount of hemoglobin and red blood cells, which results in anemia (Azim et al., 2020).

There are two types of isoenzymes: the first, ALAS-1, is found in the liver, and the second, ALAS-2, is an erythroid enzyme. The liver protein is responsible for the initiation of heme synthesis which is used by many

---

[*] Corresponding Author's Email: dbartusikaebisher@ur.edu.pl.

In: The Biochemical Guide to Enzymes
Editors: David Aebisher and Dorota Bartusik-Aebisher
ISBN: 979-8-88697-410-2
© 2023 Nova Science Publishers, Inc.

enzymes, including those related to cytochrome P450, responsible for detoxification of xenobiotics, i.e., exogenous substances, including drugs and harmful substances artificially introduced into the environment. Increased use of heme, e.g., in the case of an increased supply of xenobiotics, enhances the activity of ALA-1 synthase. ALAS-2, i.e., isoforms found in erythrocytes, does not act on xenobiotics, but takes part in the synthesis of heme used in the formation of hemoglobin (Ashorobi et al., 2020). These processes only take place in immature erythrocytes, because mature red blood cells do not contain mitochondria and the synthesis of hemoglobin and heme is inhibited (Ashorobi et al., 2020).

We owe our broad understanding of how ALAS functions in our body to many outstanding scientists and their inspiring efforts. Early work on the identification of heme precursors was performed in the Shemin and Neuberger laboratories in 1945. Shemin's focus at the time was on the study of glycine, and he discovered that it was used in the synthesis of the four heme pyrrole rings. A few years later, it was discovered that ALA is a key precursor of heme and that the enzymatic formation of this acid can be observed in bacterial extracts. Scientists had many theories about the functioning of ALAS, but ultimately, molecular cloning technology allowed them to isolate large amounts of recombinant ALAS and obtain very detailed information on the structure of this enzyme (Balwani et al., 2020). It made it possible to obtain an accurate model of the ALAS active center. At that time, the presence of a lysine residue in the active site was discovered, which covalently binds the cofactor in the absence of a substrate. Initial kinetic studies revealed that ALA release determined the rate of the reaction cycle, and that conformational changes were associated with substrate binding and product release.

ALA synthase takes part indirectly in the synthesis of porphyrins in non-plant eukaryotes, because it catalyses the condensation of glycine and succinyl-CoA, which leads to the production of 5-aminolevulinic acid (ALA) in the mitochondrial matrix, which in turn is a precursor to porphyrin synthesis. Aminolevulinic acid synthase belongs to a large family of enzymes in which pyridoxal 5-phosphate (PLP), the active form of vitamin B6, is used as a cofactor. The presence of this cofactor allows the enzyme to perform its catalytic functions. Many ALAS structures present in Rhodobacter capsulatus have been described, but our knowledge of eukaryotic ALAS was limited until recently (Balwani et al., 2020). Although the similarity between prokaryotic and eukaryotic ALAS is relatively high (there is 70% sequence similarity), there are also many features that distinguish these enzymes. Eukaryotic proteins contain a unique 35-60 residue C-terminal extension, which bacterial

ALAS does not. In the case of erythroid-specific ALAS2 isoforms, mutations can occur which cause two diseases. Mutations that shorten the C-terminal extension result in the production by the human body of overactive ALAS enzymes and the accumulation of the precursor hemu-protoporphyrin IX, which causes a toxic effect. Said mutation causes X-linked protoporphyria (XLPP) (de Paula Brandão et al., 2020). XLPP is a very rare disease characterized by severe cutaneous photosensitivity in childhood, characterized by tingling, burning and reddening of the skin. Additionally, excess protoporphyrin may accumulate in hepatocytes and bile ducts, causing hepatotoxicity, reduced bile formation and flow, and the development of liver failure in some patients. Multiple mutations in the ALAS2 isoforms in the extended C-terminus result in a partial loss of the enzyme's functions. The effect of this mutation is decreased heme synthesis, which can lead to X-linked sideroblastic anemia (XLSA). In this disease there is an increased number of pathological sideroblasts, i.e., erythroblasts containing large amounts of iron in their cytoplasm, which should be used for heme synthesis when the enzyme works properly. Such anemia is treated with folate and vitamin B6. After a few years, the disease can develop into leukemia (Wang et al., 2020).

Hemoglobin, which consists of heme, is synthesized in 85% in the erythrocyte precursor cells in the bone marrow and the remainder in hepatocytes, therefore ALA synthase is also present in these cells. ALA synthase occurs naturally in the human body in the form of two isoforms - ALAS1 and ALAS2. The first of them, present in the liver, is characterized by a fast metabolic turnover, and its synthesis is regulated by heme on the basis of feedback. In the presence of drugs whose metabolism entails a decrease in intracellular haem concentration, ALAS1 synthesis increases. It is believed that heme acts as a negative regulator of the synthesis of ALAS1 via an aporepressor molecule, and thus the increase in the amount of heme in cells inhibits the synthesis of this enzyme. Unlike ALAS1, ALAS2 biosynthesis, which is present in immature erythrocytes, is not feedback-regulated by heme. ALA synthase is an enzyme necessary for the proper functioning of our body. ALAS is needed to synthesize aminolevulinic acid, a precursor to the synthesis of porphyrins. Due to the fact that porphyrins are the basic component of heme, the deficiency of aminolevulinic acid synthase can lead to hemoglobin deficiency, and this in turn to anemia, i.e., anemia. Complete inhibition of ALA synthase production would result in death due to hypoxia in tissues and organs, including the brain and heart, which are particularly sensitive to oxygen deficiency.

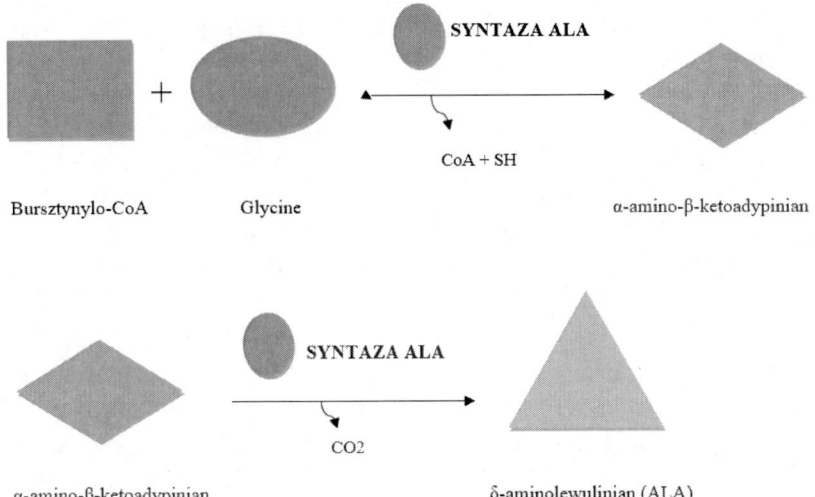

Figure 1. ALA synthase is involved in the synthesis of δ-aminolevulinate (ALA), the process is in two stages. In the first stage, under the influence of ALA synthase, condensation of succinyl-CoA takes place with glycine, as a result of which α-amino-β-keto-adipate is formed, and by-products are; CoA and SH. In the second step, α-amino-β-keto-adipate is converted into ALA by decarboxylation.

References

Ashorobi D., Chhabra A. Sideroblastic Anemia. 2021 Jul 21. In: *StatPearls* [Internet]. Treasure Island (FL): StatPearls Publishing; 2021 Jan-. PMID: 30855871.

Azim N., Gardner Q. A., Rashid N., Akhtar M. Mechanistic studies on Pyrobaculum calidifontis porphobilinogen synthase (5-aminolevulinic acid dehydratase). *Bioorg Chem.* 2019 Oct;91:103117.

Balwani M., Sardh E., Ventura P., Peiró P. A., Rees D. C., Stölzel U., Bissell D. M., Bonkovsky H. L., Windyga J., Anderson K. E., Parker C., Silver S. M., Keel S. B., Wang J. D., Stein P. E., Harper P., Vassiliou D., Wang B., Phillips J., Ivanova A., Langendonk J. G., Kauppinen R., Minder E., Horie Y., Penz C., Chen J., Liu S., Ko J. J., Sweetser M. T., Garg P., Vaishnaw A., Kim J. B., Simon A. R., Gouya L. ENVISION Investigators. Phase 3 Trial of RNAi Therapeutic Givosiran for Acute Intermittent Porphyria. *N Engl J Med.* 2020 Jun 11;382(24):2289-2301.

de Paula Brandão P. R., Titze-de-Almeida S. S., Titze-de-Almeida R. Leading RNA Interference Therapeutics Part 2: Silencing Delta-Aminolevulinic Acid Synthase 1, with a Focus on Givosiran. *Mol Diagn Ther.* 2020 Feb;24(1):61-68.

Wang Z., Gao J., Teng H., Peng J. Retracted Article: Role of aminolevulinic acid synthase 1 in doxorubicin-induced oxidative stress to the ardiomyocyte. *Naunyn Schmiedebergs Arch Pharmacol.* 2020 Nov;393(11):2231.

Chapter 12

Deiodinase

Wiktoria Kaliszewska, David Aebisher and Dorota Bartusik-Aebisher[*]
Medical College of the University of Rzeszów, Poland

Abstract

Iodothyronine deiodinases are selenoproteins acting as enzymes, characterized by the presence of selenocysteine (Sec) in the active site. In the human body, they occur in three isoforms: type I deiodinase (D1), type II deiodinase (D2) and type III deiodinase (D3). These proteins differ in their half-lives - the longest is D3 (12 hours), D1 (8 hours), and D2 is only 45 minutes.

Keywords: selenoproteins, cell metabolism, enzyme

All forms have a mass of about 30kDa and are intra-membrane enzymes. By means of the transmembrane segment, they bind to the cell membrane (D1 and D3) or the endoplasmic reticulum (D2). Thanks to this, they can more precisely fulfill their function, which is to regulate the amount of active or inactive TH. Depending on the needs of the cells, they convert T4 to T3 (or vice versa) by the reaction of iodothyronine deionization (Bianco et al., 2018).

They occur in thyrocytes as well as extra thyroid cells, the function of which depends on the hormones T3 and T4, where different isoforms may exist in different cells. Due to the fact that they regulate the presence of thyroid

---

[*] Corresponding Author's Email: dbartusikaebisher@ur.edu.pl.

In: The Biochemical Guide to Enzymes
Editors: David Aebisher and Dorota Bartusik-Aebisher
ISBN: 979-8-88697-410-2
© 2023 Nova Science Publishers, Inc.

hormones, they can be assigned the same function in the body: controlling the proper growth and development, as well as cell metabolism. The discovery of deiodinases allowed for a more detailed understanding of the causes of diseases such as: thyroid dysfunction, extra-thyroid syndrome with euthyrosis, type 2 diabetes, and mood disorders. It also made it possible to obtain a new tumor marker (D1), as well as a new factor that influences embryonic development (deiodinases are present in the fetal brain from the 7th week of pregnancy) (Bianco et al., 2018).

Type 1 deiodinase. This was the first enzyme discovered to catalyze the conversion of T4 to T3. It is distinguished from other deiodinases by means of 6-n-propylthiouracil (PTU) which inhibits its activity. In humans, the protein gene (hDio1) is located on the first chromosome in the p32.3 locus. Gene transcription is activated by thyroid hormones through TR receptors. Mutation of TR receptor proteins causes a decrease in the expression of the hDio1 gene and a decreased amount of D1 in the cell - such a condition is detected in papillary neoplasm of the thyroid gland, in clear cell neoplasm of the kidney and in glial tumors of the brain. Since the expression level is increased in neoplasms of thyroid follicles and breasts, this fact is used as a marker of neoplastic cell differentiation. This gene is expressed in highly metabolized sites such as the liver, kidneys, thyroid and mammary gland. Due to the catalytic activity of type 1 deiodinase, T3 and T2 are formed (deiodination reaction at the 5 'position) or rT3 (Bianco et al., 2018).

Type 2 deiodinase. This protein is the only protein in its structure that contains two selenium atoms. The protein gene (hDio2) is located on the long arm of chromosome 14. It is involved in the uptake of T4 from the circulating blood and then converts it into T3 which remains in target cells or is returned to plasma. This protein also has the catalytic activity to convert rT3 to T2. The greatest expression of hDio2 is in the brain (especially astrocytes, glial cells and tanycites), pituitary gland, endothelial cells, brown adipose tissue, thyroid, skeletal muscle, osteoblasts, skin and placenta. It plays a major role in thermogenesis. Particularly noteworthy are the tanicites found in the ependymal lining of the III ventricle, which play a key role in the regulation of the hypothalamic - pituitary - thyroid axis. These cells adhere to the TRH-producing paraventricular nuclei. Due to the fact that they have proteins D2 and D3, they can transform thyroxine captured from plasma or cerebrospinal fluid, as a result of which they regulate the amount of TRH by means of negative feedback. In summary, tanticites are involved in a disease entity known as central hypothyroidism, a disease that is not related to the thyroid gland (Maino et al., 2018).

Type 3 deiodinase. This is the protein gene (hDio3) located on the 14th chromosome. It is the only protein out of three that does not contain introns, and the allele it is on comes from the father. The function of D3 is to convert thyroxine to T3 or rT3 and then to T2. In adults, D3 expression is limited - it is active only in brain neurons, skin, endothelial cells, uterus and placenta. This enzyme also has a protective function against the effect of too much TSH surge. Thyrotropin stimulates the thyroid gland to produce T4 and T3, which penetrate the cell membrane of target cells and combine with the nuclear receptor, causing gene expression. This leads to the activation of proteins that enhance the metabolism of carbohydrates and fats, which increases the metabolic rate and oxygen consumption. In the liver, glycogenolysis and gluconeogenesis are enhanced, and in the heart, the activation of adrenergic receptors - the sympathetic nervous system. In pathological conditions, e.g., extreme starvation, myocardial infarction or sepsis, hDio3 is "neo-expressed." The process occurs in the liver, heart, and skeletal muscle - tissues where the gene is normally inactive. The factor inducing this mechanism is the state of hypoxia (hypoxia). As a result of its action, triiodothyronine is deactivated by the enzyme D3. This mechanism protects against deterioration of the pathological condition and, in extreme cases, against death (Marsan et al., 2020).

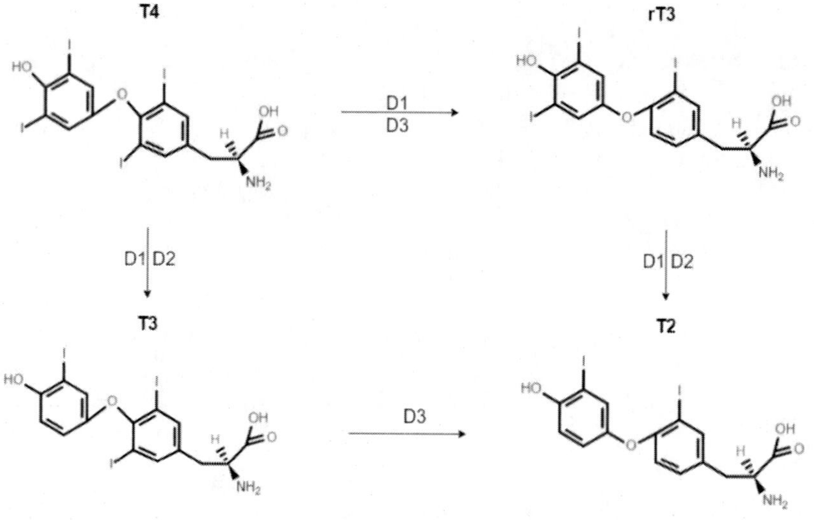

Figure 1. Transformation of thyroid hormones catalyzed by deiodinases.

Deiodinases D1 is responsible for the peripheral regulation of thyroid hormone metabolism, it remains at a low level, conditioning a low level of T3. It can be assumed that this action is aimed at limiting thermogenesis and energy consumption during anabolic processes, which must be intensified in a growing organism. D2 and D3 have the highest expression in the brain and pituitary gland of the fetus, which helps to maintain high levels of T3 in these organs. Type 2 diabetes mellitus is a multi-gene disease that may be preceded by symptoms such as such as: insulin resistance or impaired insulin secretion. Thyroid hormones and D2 and D3 play a key role in the maturation of thyroid beta cells by regulating transcription factors. They also show different expression depending on age and the development of glucose sensitivity. In people with inactivation of the D3 gene, glucose-stimulated insulin secretion is impaired (cells are intolerant to it), but insulin sensitivity is not altered. The change in the protein D2 gene (which occurs in 15% of the population) causes a slower rate of glucose degradation and insulin resistance. This fact resulted in a 10% probability of developing type 2 diabetes. Mice lacking protein showed a significant increase in body weight and fatty liver. This polymorphism has also been associated with central obesity and hypertension. These symptoms occur in the metabolic syndrome, which can also be the initial stage leading to diabetes (Sullivan et al., 2019).

References

Bianco A C, da Conceição R R. The Deiodinase Trio and Thyroid Hormone Signaling. *Methods Mol. Biol.* 2018;1801:67-83.

Bianco A C, Kim B S. Pathophysiological relevance of deiodinase polymorphism. *Curr. Opin. Endocrinol. Diabetes Obes.* 2018 Oct;25(5):341-346.

Maino F, Cantara S, Forleo R, Pilli T, Castagna M G. Clinical significance of type 2 iodothyronine deiodinase polymorphism. *Expert Rev. Endocrinol. Metab.* 2018 Sep;13(5):273-277.

Marsan E S, Bayse C A. A Halogen Bonding Perspective on Iodothyronine Deiodinase Activity. *Molecules.* 2020 Mar 14;25(6):1328.

Sullivan SA. Hypothyroidism in Pregnancy. *Clin. Obstet. Gynecol.* 2019 Jun;62(2):308-319.

Chapter 13

Phenylalanine Hydroxylase

Weronika Pawelic, David Aebisher and Dorota Bartusik-Aebisher[*]
Medical College of The University of Rzeszów, Poland

Abstract

Phenylalanine hydroxylase is a liver enzyme protein that belongs to the class of hydrolases, which means that they break chemical bonds when exposed to water. The mentioned enzyme catalyses the reaction of conversion of one amino acid (L-phenylalanine) into another (L-tyrosine). More precisely, this reaction consists in the addition of the -OH group to the carbon at the 4-position of the tyrosine.

Keywords: phenylalanine hydroxylase, exogenous, enzyme

The presence of iron ions, which act as a cofactor in this reaction, is also necessary. Phenylalanine hydroxylase, consisting of 452 amino acids, is composed of four identical subunits linked with iron atoms, which is necessary for the proper activity of this enzyme. However, each of the four subunits consists of three regulatory protein domains responsible for catalytic activity and one responsible for tetramer formation. The discussed hydroxylase, or rather, mutations that take place in the gene coding for this enzyme lead to the autosomal recessive disease, phenylketonuria. It is manifested by the accumulation (over 2 mg / dl in the blood) of this exogenous (thanks to which

---

[*] Corresponding Author's Email: dbartusikaebisher@ur.edu.pl.

In: The Biochemical Guide to Enzymes
Editors: David Aebisher and Dorota Bartusik-Aebisher
ISBN: 979-8-88697-410-2
© 2023 Nova Science Publishers, Inc.

it is possible to limit its participation in the diet and thus alleviate the course of the disease) amino acid in the blood and body tissues. It leads mainly to neurological disorders, mental and motor impairment. Interestingly, it is reported that in the first year of life, each subsequent week of a newborn child lowers his IQ by an average of one to two. Children with this disease have a characteristic (mousy) sweat odor (Kure et al., 2019).

The said enzyme is encoded by the PAH gene, which is located on the 12th chromosome (region q22-24.1). Approximately 420 mutations in the gene for this enzyme have been identified. About 60% are point mutations that reduce the activity of this enzyme. The remaining 40% are mutations of a different type, which lead to the formation of inactive forms of the enzyme. The molecular study analyzing the exons of the PAH gene showed that in the classic form of phenylketonuria, the R408W mutation is most often present in the Polish population in over 50% of the population, followed by the IVS10 mutation, c.1066-11g-a, which accounts for about 5%, IVS12 - c.1315 + 1g-aw 3.9% and R158Q accounting for about 3.5%. The PAH gene contains 13 exons and a fairly large intron span. The largest is over 23 kb. The disease caused by mutations in the PAH gene leads to phenylketonuria, which is a hereditary disease, with the exception of less than 0.1% of cases where the mutation is spontaneous, not inherited from the parents. Every 46th person in Poland is a carrier of the mutated phenylketonuria gene. This means that in a family where both parents are carriers, there is a 25% probability of giving birth to a sick child. Depending on the concentration of phenylalanine, 3 forms of the disease are distinguished: classic phenylketonuria, mild phenylketonuria, mild hyperphenylalaninemia (HPA) (Regier et al., 2021).

The incidence of this disease in the European population it is 1/10000, while in the Polish population it is 1/5000. The most common classic phenotype found in western and central-eastern Europe is a more difficult to treat. In the south, the milder phenotype of the disease is much more common. Early detection of the disease is extremely important, thanks to which it is possible to quickly prevent symptoms that are sometimes severe, such as psychomotor difficulties, decreased muscle tension, IQ drop, hyperactivity, convulsions, vomiting, damage to the central nervous system, and athetosis. For this purpose, a test is performed that measures the concentration of phenylalanine in the blood, usually taken from the child's heel. For the result of this test to be reliable, it is necessary to feed the baby with mother's milk or modified milk beforehand. Despite the rapid progress in the molecular diagnostics of the basis of phenylketonuria, so far, no effective form of therapy has been found that would allow a cure (Rink et al., 2020).

Therefore, the solution remains a properly implemented correct diet that eliminates products rich in the amino acid - phenylalanine. It is related excluding products such as meat, dairy, fish, pulses, etc., in the diet of healthy people they are a source of protein. For patients, special preparations are available that contain amino acids that provide about 80-85% of the supply of protein and the amino acid-tyrosine. The main assumption of the diet in PKU is to maintain the concentration of phenylalanine in the blood at the recommended values, which at the age of 0-12 years are 120-360 µmol / 1 (2-6 mg / dl), over 12 years old 120-600 µmol / 1 (2-10 mg / dl), and for women planning pregnancy and pregnant women 120-360 µmol / 1 (2-6 mg / dl) and ensuring the supply of energy and nutrients enabling growth and prevention of shortages. The only naturally occurring protein with small amounts of phenylalanine is glycomacropeptides, which are found naturally in whey. They are much better absorbed than synthetic preparations containing mixtures of amino acids. In patients with mild phenylketonuria, the use of sapropterin hydrochloride appears to be an effective therapy in which the body's ability to convert phenylalanine into tyrosine is increased. The introduction of this substance seems to be a method to expand the diet. Both therapy with glycomacropeptides and sapropterin hydrochloride is not reimbursed in Poland (Stone et al., 2021).

Taking into account the influence of the abnormally functioning liver enzyme on the body (although its presence was also confirmed in human kidneys, heart muscle, small intestine, lungs and spleen. The antigen content in the tested tissue extracts was (µg / g): liver - 1500-1900, kidney - 300-575, brain - 20-40, heart muscle - 85-105, lung - 40-125, small intestine - 45 -70, spleen - 0-12). Phenylalanine hydroxylase can be assessed as important and crucial for proper development and life is a protein. It is surprising how a malfunction of one enzyme that determines the conversion of tyrosine into phenylalanine affects the intellectual abilities of a person. It has been found that during the first year of life, each subsequent week of a newborn baby lowers his IQ by an average of one to two when phenylketonuria is present. Each of the search for an effective drug so far has resulted in a lack of answers on how to treat this genetically conditioned disease. However, with each subsequent attempt, the knowledge about the dysfunction of this protein expands, which may lead to an effective treatment in the future. So far, the only method is to follow a strict low-protein diet. Phenylalanine hydroxylase (PH) activity was discovered in the liver in as early as 7–12-week-old human embryos. The efficacy of therapy in treating patients with phenylketonuria during infancy is very good. However, it worsens significantly in the following years, what is more, mutations unrelated to family origin are more frequent then, therefore it is extremely important to detect phenylalanine hydroxylase dysfunction early in life (van Wegberg et al., 2017).

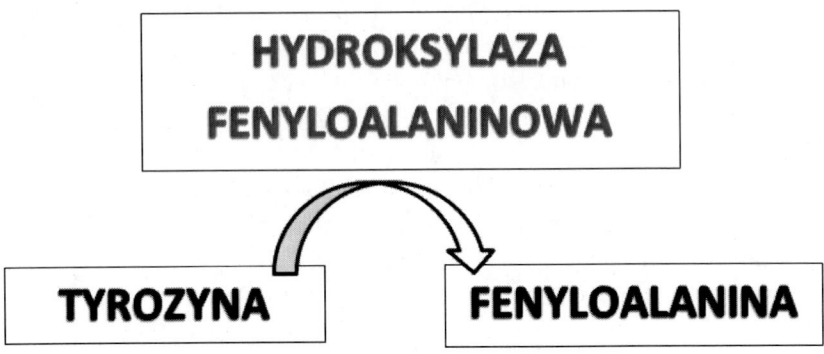

Figure 1. Diagram of the transformation conditioned by phenylalanine hydroxylase, whereby tyrosine is transformed into phenylalanine.

References

Kure S, Shintaku H. Tetrahydrobipterin-responsive phenylalanine hydroxylase deficiency. *J. Hum. Genet.* 2019 Feb;64(2):67-71.

Regier D S, Greene C L. Phenylalanine Hydroxylase Deficiency. 2000 Jan 10 [updated 2017 Jan 5]. In: Adam MP, Ardinger HH, Pagon RA, Wallace SE, Bean LJH, Gripp KW, Mirzaa GM, Amemiya A, editors. *GeneReviews®* [Internet]. Seattle (WA): University of Washington, Seattle; 1993–2021. PMID: 20301677.

Rink B, Dukhovny S, Timofrev J. Management of Women With Phenylalanine Hydroxylase Deficiency (Phenylketonuria): ACOG Committee Opinion, Number 802. *Obstet. Gynecol.* 2020 Apr;135(4):e167-e170.

Stone WL, Basit H, Los E. Phenylketonuria. 2021 Aug 11. In: StatPearls [Internet]. Treasure Island (FL): *StatPearls Publishing*; 2021 Jan–. PMID: 30570999.

van Wegberg A M J, MacDonald A, Ahring K, Bélanger-Quintana A, Blau N, Bosch A M, Burlina A, Campistol J, Feillet F, Giżewska M, Huijbregts S C, Kearney S, Leuzzi V, Maillot F, Muntau A C, van Rijn M, Trefz F, Walter J H, van Spronsen F J. The complete European guidelines on phenylketonuria: diagnosis and treatment. *Orphanet. J. Rare Dis.* 2017 Oct 12;12(1):162.

Chapter 14

Nitrogenase (Flavodoxin)

Martyna Paśko, David Aebisher and Dorota Bartusik-Aebisher*
Medical College of the University of Rzeszów, Poland

Abstract

Nitrogenases (nitrases) are enzymatic complexes belonging to the group of metalloenzymes, found in archaea and bacteria - aerobic and anaerobic, autotrophs, heterotrophs and cyanobacteria. They require the presence of magnesium ions, less often vanadium. They are not found in any eukaryotic organisms.

Keywords: nitrogenases, enzymes, biochemistry

Nitrogenases, their main function is the assimilation of atmospheric nitrogen (reduction) from the N2 form to the ammonia form, NH3, which is bioavailable for many organisms. They belong to the so-called flavodoxins due to their functional interchangeability with ferredoxins. They are complex proteins consisting of two components: dinitrogenase, MoFe cofactor, i.e., molybdenoferredoxin (larger unit approx. 180-270 kD), and dinitrogenase reductase (smaller unit 55-65 kD) (Figure 1). It should be emphasized that nitrogenases isolated from different sources have similar physicochemical and catalytic properties. Both phylogenetic evidence and geochemistry suggest that nitrogenases are an old family of enzymes; phylogenetics estimates the

---

* Corresponding Author's Email: dbartusikaebisher@ur.edu.pl.

In: The Biochemical Guide to Enzymes
Editors: David Aebisher and Dorota Bartusik-Aebisher
ISBN: 979-8-88697-410-2
© 2023 Nova Science Publishers, Inc.

evolution of nitrates at around 1.5-2.2 billion years, while the oldest biosignatures provide the activity of these enzymes dates back about 3.2 billion years (Badalyan et al., 2019).

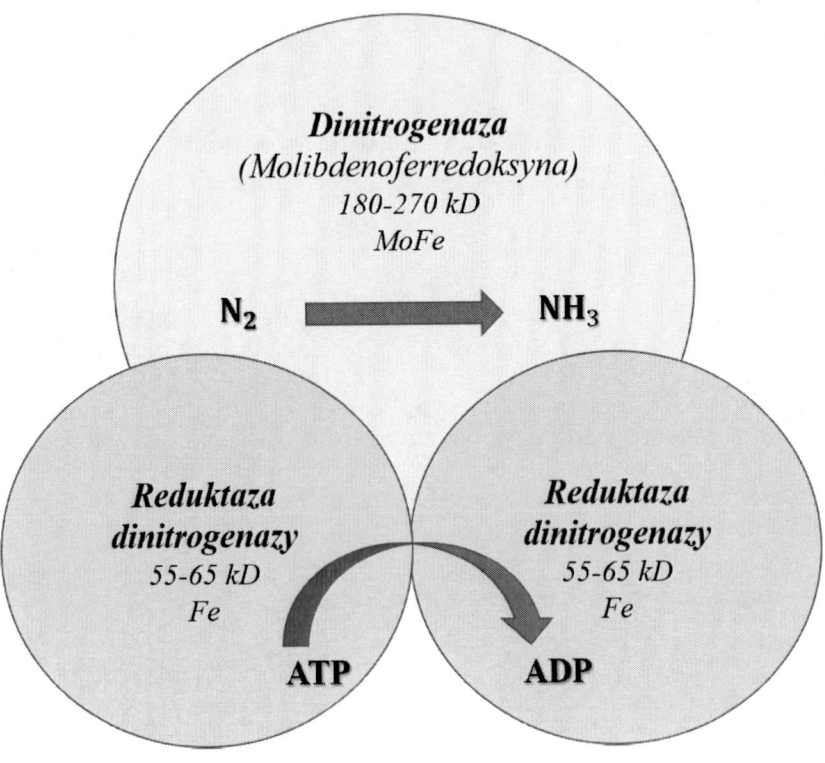

Figure 1. Diagram of the structure of the nitrogenase complex.

It took almost 80 years since the discovery of nitrogen fixation before Bulen and LeComte successfully purify the first nitrogenase enzyme in 1966 and another 30 years before Rees and colleagues reported the first X-ray crystallographic structures of constituent proteins and nitrase active sites in 1992-93 - a remarkable advance in the field of nitrase research. Nitrogenase is an enzyme still under investigation. In any ecosystem, dinitrophs must respond to different environmental conditions to regulate the highly stressful nitrogen fixation process. All characterized diazotrophs regulate nitrase at the transcriptional level. The smaller kit also has a fast-acting post-translational regulation system. While there is little marked variation in the sequences and structures of the nitrases, there appears to be nearly as many nitrase regulating patterns as there are nitrogen-fixing species themselves. Perhaps the best recognized mechanism for N2 binding in the aerobic environment is related to symbiotic nitrogen fixation, in which plants provide a micro-oxygen niche in which oxygen tension is kept low by a high affinity oxygen binding protein called leghemoglobin, which is produced by the host plant (Fixen et al., 2018).

Strategic oxygen sequestration allows symbiotic dinitrophs (e.g., Rhizobium bacteria) to maintain aerobic respiration while catalysing the fixation of atmospheric nitrogen that occurs under strictly anaerobic conditions and only during periods of anaerobic growth in facultative anaerobes. Cyanobacteria, the only diazotrophic lineage that produces molecular O_2 as a product of its metabolism, has developed a number of N2 binding mechanisms. For example, non-filamentous cyanobacteria operate in a diurnal cycle, in which nitrogen fixation is increased at night, when the oxygen partial pressure has decreased due to decreased photosynthesis and increased oxygen consumption by heterotrophs inhabiting the same niche.

Numerous studies are underway to further understand the structure and function of this enzyme so that it can be reproduced in vitro. Unresolved questions concern the exact nature of the intermediates between N2 and NH3 and whether the cofactor rearranges during marketing. Resolving these ambiguities will require the consolidation of biochemistry, spectroscopy and model chemistry research. This would allow the design of new biological catalysts that exploit the ability of nitrogenase to produce ammonia (such as directing renewable electricity towards MgATP-independent N2 binding by MoFe protein). In addition, a number of natural and unnatural electron donors are being studied (Poudel et al., 2018).

In the case of natural electron donors, it is still necessary to improve the efficiency of the electron transfer systems in non-nitrous hosts to support higher nitrase activity. The symbiotic relationship between legumes and Rhizobium is agronomic and limits the use of nitrogen fertilizers. The long-term use of nitrogen bio-fertilization in the agricultural field is a promising approach to develop and meet the needs of crops for N_2 without causing any threat to the environment. It is difficult to use the nitrase complex in laboratories because it must be highly stable and oxygen tolerant. In addition, while attempts to heterologously express nitrase complexes in plant cells remain a challenge, not least because about half of the enzyme's subunits are insoluble, many methods have been developed to increase the solubility of proteins presenting potential solutions. As a result, it is expected that in the future the production of zero-emission food will be promoted by discontinuing the use of chemical fertilizers (Pence et al., 2017).

Nitrogen is the main limiting nutrient in high-latitude ecosystems. Biological nitrogen fixation (BNF) by microorganisms associated with cryptogamous mites, such as cyanobacteria and bryophytes, is an important source of new reactive nitrogen in virgin ecosystems. Of course, BNF is catalyzed by the enzyme nitrase for which three isoforms with molybdenum, vanadium and iron have been described. Low availability of molybdenum on land has been shown to reduce BNF in many ecosystems, from tropical forests to the Arctic tundra. Hence, it can be concluded that nitrase is the basis of almost every ecosystem and thanks to it we can observe such great biodiversity and unique flora and fauna on every continent (Pence et al., 2017).

In recent years, our understanding of the nitrogen cycle has been broadened to include both new nitrogen fixing environments and new organisms performing this task. Although Mo-Nitrogenases have been well researched in the environment, knowledge of alternative nitrases is still scarce with little information about their distribution in nature. The body's "choice" to use one nitrase as opposed to another is complex and more research will be needed to do so understand this phenomenon. In the last 5 years, there has been significant progress in understanding the mechanism of nitrogenase reduction of substrates, in particular in the identification of hydride and nitrogen containing intermediates using pulsed EPR 205 methods and in the crystallographic characterization of nitrase forms (Segal et al., 2017).

Advances in working with small molecule catalysts complement the multidisciplinary approach to the enzyme system by providing an increasingly detailed view of the interior of the still mysterious nitrite. Also, understanding the nature of an active catalyst and its potential deactivation pathways can facilitate the development of new and improved catalytic systems. For non-physiological nitrase substrates, an important question is understanding how the molecular form of nitrogen is hydrogenated. Is this happening through migratory insertion of hydride species associated with the FeMo cofactor? Is there any significant structural rearrangement that plays an important role during bonding or hydrogenation? The answers to these questions will take our understanding of the mechanism to a higher level. The direct application of nitrase or its cofactors to raw material production or CO_2 sequestration on a useful scale is likely to be an insurmountable challenge given the lack of resistance of the enzymes and their associated cofactors (Segal et al., 2017).

References

Badalyan A, Yang Z Y, Hu B, Luo J, Hu M, Liu T L, Seefeldt L C. An Efficient Viologen-Based Electron Donor to Nitrogenase. *Biochemistry*. 2019 Nov 19;58(46):4590-4595.

Fixen K R, Pal Chowdhury N, Martinez-Perez M, Poudel S, Boyd E S, Harwood C S. The path of electron transfer to nitrogenase in a phototrophic alpha-proteobacterium. *Environ. Microbiol*. 2018 Jul;20(7):2500-2508.

Pence N, Tokmina-Lukaszewska M, Yang Z Y, Ledbetter R N, Seefeldt L C, Bothner B, Peters J W. Unraveling the interactions of the physiological reductant flavodoxin with the different conformations of the Fe protein in the nitrogenase cycle. *J. Biol. Chem.* 2017 Sep 22;292(38):15661-15669.

Poudel S, Colman D R, Fixen K R, Ledbetter R N, Zheng Y, Pence N, Seefeldt L C, Peters J W, Harwood C S, Boyd E S. Electron Transfer to Nitrogenase in Different Genomic and Metabolic Backgrounds. *J. Bacteriol.* 2018 Apr 24;200(10):e00757-17.

Segal H M, Spatzal T, Hill M G, Udit A K, Rees D C. Electrochemical and structural characterization of Azotobacter vinelandii flavodoxin II. *Protein Sci.* 2017 Oct;26(10):1984-1993.

Chapter 15

DMSO Reductase

Kacper Wygonik, David Aebisher and Dorota Bartusik-Aebisher[*]
Medical College of the University of Rzeszów, Poland

Abstract

Dimethyl sulfoxide reductases (DMSO) are a family of very old evolutionarily mononuclear molybdoenzymes. They can only be detected in prokaryotes (Bacteria and Archaea domains).

Keywords: Dimethyl sulfoxide reductases, metabolism, enzyme, biochemistry

Based on the research on the evolution of the biogeochemical arsenic cycle, a phylogenetic tree was created with over 1,550 representatives of the DMSO reductase family. Their cofactor is bis-molybdopterynoguanine dinucleotide (Bis-MGD). Molybdenum is a transition metal. It plays an essential role in the metabolism of organisms belonging to all three domains (Archea, Bacteria, Eukaryota), but in its free state it cannot be absorbed by living organisms. The biologically available form of molybdenum is molybdate oxyanion (MoO_{24}) (Barnum et al. 2020). The function of DMSO reductases is not only to catalyze redox reactions, but they are also involved in the processes of hydroxylation and oxygen transfer. They play an important role in the biogeochemical cycles of the biogenic elements carbon, nitrogen and sulfur, as well as other elements

---

[*] Corresponding Author's Email: dbartusikaebisher@ur.edu.pl

In: The Biochemical Guide to Enzymes
Editors: David Aebisher and Dorota Bartusik-Aebisher
ISBN: 979-8-88697-410-2
© 2023 Nova Science Publishers, Inc.

such as arsenic, selenium and possibly also antimony. Some enzymes belonging to the DMSO reductase family show the ability to reduce methionine sulfoxide (MetO) in free form or incorporated into the protein structure. This is a very important task, because the oxidation of methionine residues can change the properties of proteins in many ways - it can have a negative impact on the stability of their structure as well as the activity of enzymes and the ability to interprotein interactions. It is also believed that DMSO reductases were one of the most important elements in the first chains of anaerobic respiration, which would indicate that these enzymes were involved in the process of creating life on Earth (Dong et al. 2017).

Contrary to other families belonging to enzymes containing molybdenum or tungsten, DMSO reductases are present only in prokaryotes, what is more, they are present in a large part of these organisms. Over 90% of organisms that use molybdenum in their metabolism are equipped with enzymes belonging to the family of DMSO reductases. On this basis, it can be concluded that this family is the most common group of molybdoenzymes among prokaryotes (Miralles-Robledillo et al. 2019). Examples of representatives of this family are summarized in Table 1.

Table 1. Examples of enzymes belonging to the DMSO reductase family

Name of the enzyme	Type	Catalyzed reaction	Organism
DMSO reductase	III	$(CH_3)_2SO + 2e^- + + 2H^+ \rightarrow (CH_3)_2S + + H_2O$	*Rhodobacter sphaeroides, Rhodobacter capsulatus*
Respiratory nitrate reductase (Nar)	II	$NO_3^- + 2e^- + 2H^+ \rightarrow \rightarrow NO_2^- + H_2O$	*Escherichia coli*
Periplasmic Nitrate Reductase (Nap)	I	$NO_3^- + 2e^- + 2H^+ \rightarrow \rightarrow NO_2^- + H_2O$	*Desulfovibrio desulfuricans*
TMAO reductase	III	$(CH_3)_3NO + 2e^- + + 2H^+ \rightarrow (CH_3)_3N + + H_2O$	*Shewanella massilia*
Formate dehydrogenase-H	I	$HCOOH \rightarrow CO_2 + + 2e^- + 2H^+$	*Escherichia coli*
N-formate dehydrogenase	I	$HCOOH \rightarrow CO_2 + + 2e^- + 2H^+$	*Escherichia coli*
Arsenine oxidase	-	$AsO_2^- + H_2O \rightarrow \rightarrow AsO_3^- + 2e^- + 2H^+$	*Alcaigenes faecalis*

Based on the differences in the structure of the molybdenum binding site, DMSO reductases are divided into three subfamilies (types) denoted by Roman numerals:

- Subfamily I: periplasmic nitrate reductase (Nap) and formate dehydrogenase (Fdh)
- Subfamily II: respiratory nitrate reductase (Nar) and ethylbenzene dehydrogenase (EBDH)
- Subfamily III: DMSO reductase, TMAO reductase

Arsenic oxidase cannot be included in either of these subfamilies due to the side chain of the amino acid attached to the molybdenum atom (see below) (Miralles-Robledillo et al. 2019).

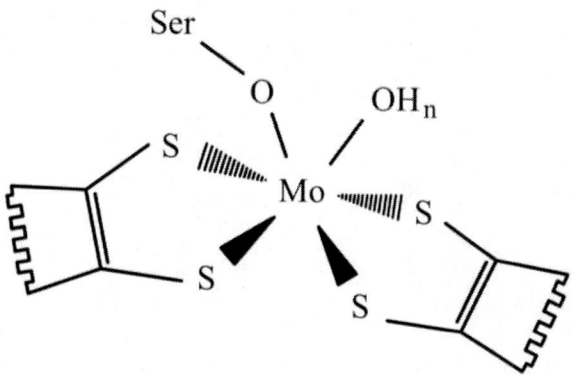

Figure 1. Structure of the DMSO reductase active site.

The molybdenum cofactor (Moco) in all members of the DMSO reductase family is bound by four sulfur atoms that attach to two nucleoside-linked pyranopterin molecules (cytidine or guanosine). In addition, the molybdenum atom is coordinated by an inorganic ion (selenium, oxygen or sulfur). Another molybdenum ligand found in nearly all DMSO reductases is an amino acid side chain, which can be derived from aspartic acid, serine, cysteine, or selenocysteine. It is because of the latter type of ligand that among these enzymes we distinguish the above-mentioned subfamilies: I (cysteine or selenocysteine), II (aspartate), III (serine) (Mintmier et al. 2021).

In Rhodobacter spheroides, the protein part consists of 780 amino acid residues, in Escherichia coli, DMSOR is a membrane protein. The secondary structure of DMSO reductase (as the actual enzyme) is represented by the α and β structures. The molecule of this protein can be divided into four domains, of which only one does not interact directly with the cofactor. All the domains gathered around the cofactor form a slightly elongated 75 by 55 by 65 Å molecule (Mintmier et al. 2021).

As the name suggests, DMSO reductase (EC 1.8.5.3), or dimethyl sulfoxide reductase, catalyzes the reduction of this compound to dimethyl sulfide (DMS).

$$(CH_3)_2S{=}{=}O + 2H^+ + 2e^- \rightarrow (CH_3)_2S + H_2O$$

This reaction is crucial in the processes of anaerobic respiration in prokaryotes. In its course, the oxygen atom is detached from the DMSO molecule and two electrons go in its place, thanks to which DMSO becomes the final electron acceptor in the process of oxidative phosphorylation in anaerobes. The detached oxygen atom combines with two hydrogen cations to form a water molecule (Wells et al. 2020).

Basically, the process of making Moco can be described in four points:

(1) Formation of cyclic pyranopterin monophosphate (cPMP);
(2) Conversion of cPMP into molybdopterin (MPT);
(3) Insertion of a molybdenum atom - creation of Moco;
(4) Additional Moco modifications by attachment GMP (MGD) or CMP (MCD).

A total of 16 genes located in 6 different loci are involved in the biosynthesis of the molybdenum cofactor: moa, mob, potency, mod, moe and mog. The protein portion of the enzyme is encoded by the dmsABC gene.

Based on the data presented on the PubMed website (as of November 2021), after entering the phrase "DMSO reductase" in the search engine, it can be concluded that this protein is not very popular. Most publications appeared in 2001, when their number was 21, while in 2020 it was only 11 items. This is a relatively small number considering the fact that in recent years the number of publications on insulin has reached 18,000 annually (Wells et al. 2020).

DMSO reductase is the "title" and therefore the most important member of the enzyme family of the same name. The tasks of this family of enzymes include: (1) catalysis of redox reactions; (2) these enzymes participate in the cycles of many elements (biogenic as well as others); (3) control of the correct structure and function of other enzymes and proteins. DMSO reductase itself, understood as the actual enzyme, plays a key role in anaerobic prokaryotes. Its presence enables the process of anaerobic respiration (more precisely the respiratory chain), reducing dimethylsulfoxide, which acts as the final electron pair acceptor - it is the last link in the respiratory chain.

The production of biologically fully functional reductase requires the availability of molybdenum in an available form for those anaerobes that use this element to synthesize the enzyme's cofactor. In the case of a deficiency of molybdenum, significant disturbances in the process of anaerobic respiration may occur. Apart from molybdenum and typical elements included in the structure of proteins (carbon, hydrogen, oxygen, nitrogen, sulfur), the structure of this reductase (more specifically the molybdenum cofactor) may include selenium (in selenocysteine or as an inorganic ion coordinating the molybdenum atom).

While the enzyme itself is not very popular among researchers and seems not to be very important in medicine, its substrate, i.e., dimethyl sulfoxide (DMSO), arouses much greater interest in the scientific community. It is used in veterinary and alternative medicine. This compound has recently been shown to have a positive effect on the regeneration of peripheral nerve fibers in rats.

References

Barnum TP, Coates JD. An uncharacterized clade in the DMSO reductase family of molybdenum oxidoreductases is a new type of chlorate reductase. *Environ Microbiol Rep.* 2020;12(5):534-539.

Dong G, Ryde U. Effect of the protein ligand in DMSO reductase studied by computational methods. *J Inorg Biochem.* 2017;171:45-51.

Mintmier B, McGarry JM, Bain DJ, Basu P. Kinetic consequences of the endogenous ligand to molybdenum in the DMSO reductase family: a case study with periplasmic nitrate reductase. *J Biol Inorg Chem.* 2021;26(1):13-28.

Miralles-Robledillo JM, Torregrosa-Crespo J, Martínez-Espinosa RM, Pire C. DMSO Reductase Family: Phylogenetics and Applications of Extremophiles. *Int J Mol Sci.* 2019;20(13):3349.

Wells M, Kanmanii NJ, Al Zadjali AM, Janecka JE, Basu P, Oremland RS, Stolz JF. Methane, arsenic, selenium and the origins of the DMSO reductase family. *Sci Rep.* 2020;10(1):10946.

Chapter 16

Choline Acetyltransferase

Maria Słobodzian, David Aebisher and Dorota Bartusik-Aebisher[*]
Medical College of The University of Rzeszów, Poland

Abstract

Choline acetyltransferase (ChAT) is an enzyme produced in the central nervous system. This protein is synthesized in the body of cholinergic nerve cells, from where it is then transported to the axonal endings. Most probably, this transport takes place on a fast and slow road.

Keywords: choline acetyltransferase, renal terminals, enzymes, biochemistry

Two forms of ChAT can be distinguished in the renal terminals, which include: the soluble form, which is the main form of occurrence, and the non-ionic form, which remains bound to the neuron membrane. The importance of the non-ionic form of the enzyme is unknown. In axonal endings, choline acetyltransferase participates in the synthesis of the neurotransmitter, acetylcholine, which is of great importance in the proper functioning of the nervous system. The role of ChAT in this process is to catalyze the transfer of the acetyl group from acetyl-CoA to choline. Choline acetyltransferase is a specific indicator of the control of the functional state of cholinergic neurons in both the central and peripheral nervous systems (Zhang et al., 2020).

---

[*] Corresponding Author's Email: dbartusikaebisher@ur.edu.pl.

In: The Biochemical Guide to Enzymes
Editors: David Aebisher and Dorota Bartusik-Aebisher
ISBN: 979-8-88697-410-2
© 2023 Nova Science Publishers, Inc.

A decrease in ChAT activity indicates damage to the cholinergic neurons. Disorders related to the expression of ChAT may occur in many neurodegenerative diseases, e.g., in Alzheimer's disease, schizophrenia or amyotrophic lateral sclerosis. Structurally, ChAT is a single-stranded globular protein. It belongs to the family of choline-carnitine acetyltransferases, which also includes carnitine acetyltransferase, carnitine palmitotransferase and carnitine octanol transferase. These are enzymes involved in the transformation of fatty acids and the maintenance of an appropriate amount of acetyl-CoA. Multiple transcripts are produced from a single ChAT gene that mainly encode a form of the enzyme with a molecular weight of B68 kDa. Forms weighing 82 kDa can also be formed in primates (Zhang et al., 2020).

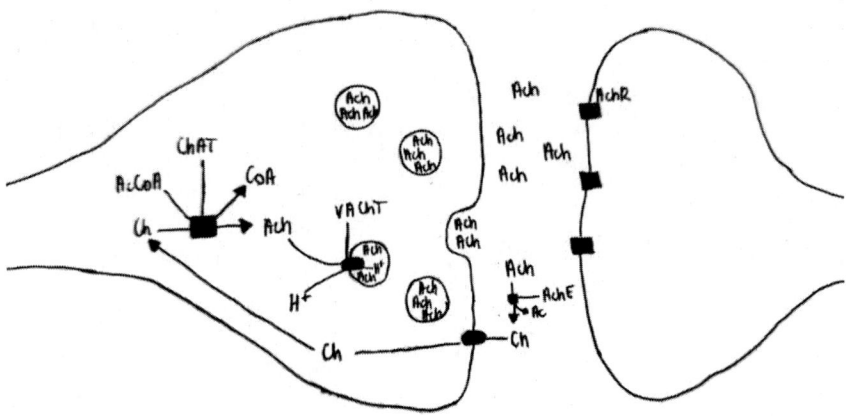

Figure 1. The role of ChAT in the synthesis of acetylcholine at the synapse.

There are several variations of ChAT mRNA - R, N, M types. They are transcribed from different promoter regions. In humans, M-type mRNA can produce both small and large forms of ChAT, while R and N mRNAs generate a small form of this protein. The first ChAT intron is the open reading frame that codes for the vesicular acetylcholine transporter (VAChT) responsible for the transport of acetylcholine to the synaptic vesicles.

Regulating the activity of ChAT has a significant influence on the development and activity of cholinergic neurons. The regulation of this enzyme probably involves mainly temporal and spatial control of transcription. Expression can also be regulated at the level of RNA transfer or processing. The regulation of ChAT can occur at the post-translational level by phosphorylation, non-covalent modification. There is also regulation of

ChAT at various levels, with both positive and negative feedback (Toan et al., 2021).

Post-translational phosphorylation is a common mechanism for regulating protein function. ChAT phosphorylation is important in cholinergic transmission. It was initially argued that ChAT was phosphorylated on the Ser residue by CaM or PK-C kinase (Bruce and Hersh 1989). It turned out, however, by analyzing the amino acid sequence of ChAT, that this enzyme has phosphorylation sites for more protein kinases. Phosphorylation of the residues in ChAT by one or more kinases may be in the form of a hierarchical sequence. ChAT phosphorylation leads to changes in its catalytic activity, subcellular localization, or interaction with other cells. Disturbances in the activity or distribution of protein kinases may change the function of choline acetyltransferase and the consequences of Ach synthesis and interneuronal communication. This is important in the course of many diseases of the nervous system (Toan et al., 2021).

ChAT is involved in the aging of the brain. With age, the activity of this enzyme decreases, and thus the synthesis of acetylcholine and transmission in cholinergic neurons are weakened. Biochemical methods for measuring ChAT activity have long been used as a marker of normal function or disturbance of the cholinergic pathways. The decrease in the activity of this enzyme is seen in Alzheimer's disease. Typical for this disease is a reduction in the concentration of choline acetyltransferase, proportional to the damage to the cholinergic system and thus the severity of the disease. It has been shown that AD can be alleviated by supplementation with exogenous ChAT. This compensates for the decreased amount of acetylcholine and thus alleviates memory and cognitive impairment. ChAT gene polymorphism plays a role in the risk of developing Alzheimer's disease (Han et al., 2017).

ChAT also correlates with schizophrenia. A study was carried out in which it was shown that the concentration of choline acetyltransferase in patients suffering from schizophrenia was 46% lower than in healthy subjects. There is a significant decrease in the activity of ChAT in the nucleus accumbens and in the tip of the pons in schizophrenics. Increased levels of choline have also been identified, indicating variable ChAT acetyltransferase activity in these individuals. ChAT is suspected to have a genetic influence on susceptibility to the disease and treatment of schizophrenia. Another disease associated with choline acetyltransferase disorders is congenital myasthenic syndrome. Mutations in ChAT impair the production of acetylcholine, leading to the onset of this disease. After analyzing seven ChAT mutations, it was concluded that ChAT mutations, which result in decreased enzyme expression and severe kinetic effects, are associated with the most severe disease phenotypes. In one study performed in mice, it was shown that the reduction of acetylcholine synthesis, caused by the reduction of ChAT, may cause dysfunction of the salivary glands (Cox et al., 2019).

The substances increasing the expression of ChAT include estrogens, the receptors of which are found, inter alia, in the spinal cord. Studies in mice have shown that 17B-estradiol can increase the expression of ChAT in the spinal cord. Genistein can stimulate ChAT expression by activating estrogen receptors. One study showed that ChAT activity is increased when a neuron is in a depolarized state. Vitamin D3 derivatives also increase the activity of choline acetyltransferase (Abicht et al., 2021).

In conclusion, choline acetyltransferase is a very important enzyme in the nervous system. By its function, which is participation in synthesis. acetylcholine, plays a significant role in nerve conduction in cholinergic neurons. The level of ChAT expression may be an important indicator to pay attention to when examining diseases associated with disorders of the nervous system. It is worth focusing especially on the reduced activity of this enzyme, because such a disorder correlates with the diseases described above. ChAT expression is related to estrogen receptors, which may indicate their indirect role in the functioning of cholinergic neurons.

References

Abicht A, Müller JS, Lochmüller H. Congenital Myasthenic Syndromes Overview. 2003 May 9 [updated 2021 Dec 23]. In: Adam MP, Ardinger H H, Pagon R A, Wallace S E, Bean L J H, Gripp K W, Mirzaa G M, Amemiya A, editors. *GeneReviews®* [Internet]. Seattle (WA): University of Washington, Seattle; 1993–2021. PMID: 20301347.

Cox M A, Duncan G S, Lin G H Y, Steinberg B E, Yu L X, Brenner D, Buckler L N, Elia A J, Wakeham A C, Nieman B, Dominguez-Brauer C, Elford A R, Gill K T, Kubli S P, Haight J, Berger T, Ohashi P S, Tracey K J, Olofsson P S, Mak T W. Choline acetyltransferase-expressing T cells are required to control chronic viral infection. *Science.* 2019 Feb 8;363(6427):639-644.

Han B, Li X, Hao J. The cholinergic anti-inflammatory pathway: An innovative treatment strategy for neurological diseases. *Neurosci. Biobehav. Rev.* 2017 Jun;77:358-368.

Toan N K, Tai N C, Kim S A, Ahn S G. Choline Acetyltransferase Induces the Functional Regeneration of the Salivary Gland in Aging SAMP1/Kl -/- Mice. *Int. J. Mol. Sci.* 2021 Jan 2;22(1):404.

Zhang X, Lei B, Yuan Y, Zhang L, Hu L, Jin S, Kang B, Liao X, Sun W, Xu F, Zhong Y, Hu J, Qi H. Brain control of humoral immune responses amenable to behavioural modulation. *Nature.* 2020 May;581(7807):204-208.

Chapter 17

Glutaredoxins

**Radosław Starzyk, David Aebisher
and Dorota Bartusik-Aebisher***
Medical College of the University of Rzeszów, Rzeszów, Poland

Abstract

Glutaredoxins (Grxs) are small intracellular thiol enzymes that belong to reductases. They belong to the Grx system consisting of glutaredoxin, glutathione, glutathione reductase and NADPH. Glutaredoxins were discovered by Arne Holmgren, a Swedish professor of biochemistry in 1976. They were isolated from mutant Escherichia coli bacteria.

Keywords: glutaredoxins, thioredoxins, enzymes, biochemistry

At that time, thioredoxins (Trx) were the only discovered hydrogen donor for ribonucleotide reductase (RNR). It is now known that both Trx and Grx function as hydrogen donors for ribonucleotide reductase in the synthesis of deoxyribonucleotides. However, despite the same function in this process and their similarity in structure). The active center, there are many different properties of these compounds. Years of research have shown that glutaredoxins are not only a substitute in the absence of Trx as once thought but play important roles in many processes. The role of Grx in regulating development in metabolism and stress response has been confirmed. So far, 4 isoforms of human Grx have been discovered and divided according to the amount of the cysteine amino acid in the active center of the enzyme: Grx3

---

* Corresponding Author's Email: dbartusikaebisher@ur.edu.pl.

In: The Biochemical Guide to Enzymes
Editors: David Aebisher and Dorota Bartusik-Aebisher
ISBN: 979-8-88697-410-2
© 2023 Nova Science Publishers, Inc.

and Grx5 monothiols with one cysteine, Grx1 and Grx2 dithiols with two cysteines. Grx1 occurs mainly in the cytosol (Figure 1). Grx2 is encoded by one gene but occurs in three variants: Grx2a appearing in the mitochondrial matrix (Figure 2), G rx2b and Grx2c found in the cytosol of testicular cells and neoplastic cells. Grx5 is found in the mitochondria and is necessary for the biogenesis of the Fe-S group (Burns et al. 2020).

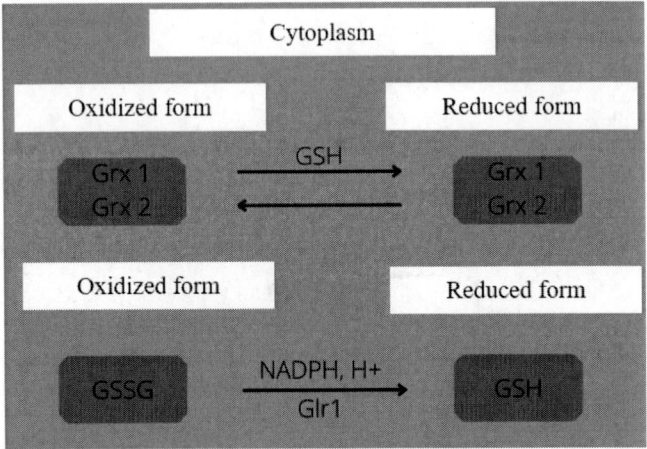

Figure 1. Oxidation and reduction of glutaredoxins in the cytosol. The reduction of Grx1 and Grx2 is due to GSH. The reduction of GSSG occurs with the participation of NADPH and H + and is catalyzed by GSH reductase (Glr1).

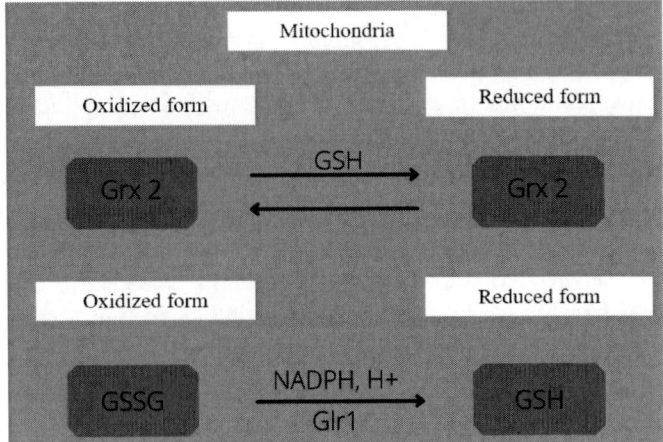

Figure 2. Oxidation and reduction of glutaredoxins in the mitochondria. Grx2 reduction is due to GSH. The reduction of GSSG occurs with the participation of NADPH and H + and is catalyzed by GSH reductase (Glr1).

Main Functions of Glutaredoxins

Glutaredoxins are necessary for the synthesis of deoxyribonucleotides from ribonucleotides by the enzyme glutathione-dependent ribonucleotide reductase (GSH). Additionally, glutaredoxins from Escherich coli bacteria have transhydrogenase activity that reduces some disulfides together with GSH reductase. Synthesis of deoxyribonucleotides can also take place through thioredoxin-dependent reactions. Glutaredoxins and thioredoxins are similar in size but function. Grx are more important for the oxidoreductase reaction, not the transferase reaction (Gorelenkova et al. 2018).

Protein thiolation by GSH was initially described as a non-specific modification induced by strong oxidative conditions that occurs during so-called oxidative stress. It is a dynamic process in which glutaredoxins have a significant regulatory function. The theory of glutathionylation as an important mechanism for oxidative signal transduction, first introduced by John J. Mieyal et al. in 1995, is supported by the increase in the discoveries of proteins believed to be modulated by glutathionylation by glutaredoxins. The so far identified proteins susceptible to glutathionylation can be divided into six different classes: cytoskeleton, metabolic glycolysis pathways, signaling pathways, involved in maintaining calcium homeostasis, antioxidant and involved in protein formation. There are two criteria that characterize the glutathionylation mechanism as a regulatory mechanism. The first is that this process only occurs with certain cysteine molecules in the protein. The second is the reversibility of the glutathionylation process allowing signal recovery. Glutaredoxins, and in particular the forms Grx1 and Grx2, catalyze deglutathionylation reactions much more efficiently than thiol transferases such as Trx or disulfide isomerases, allowing for a reversal of the signal. Despite the research being done to understand the mechanism by which the redox signal is transferred from the oxidant to special proteins and how exactly glutaredoxins are activated to carry out the glutathionylation process, many aspects remain unexplained so far (Gorelenkova et al. 2018).

The Grx3 and Grx5 glutaredoxin isoforms are evolutionarily conserved in both bacteria and mammals. They are very important in the processes of iron transport, maintaining its homeostasis and storage. Grx3 and Grx5 form an iron-sulfur complex. Recent studies have shown that both of these forms have the ability to transport iron to specific proteins and then to its binding sites. Glutaredoxins are classified based on various parameters such as amino acid sequence similarity, structure, enzymatic activity, and more. The most important, however, is the division into monothiols and dithiols due to the number of cysteines in the active site of the molecule. Monothiols (Grx3, Grx5), unlike dithiols (Grx1, Grx2), do not have the ability to deglutathylate targeted proteins, therefore their participation in redox reactions remains controversial. Interestingly, dithiol glutaredoxins may behave like monothiols, but the essence of this mechanism and the conditions under which it occurs have not yet been understood (Jacquot et al. 2019).

The two most known human glutaredoxins are Grx1 and Grx2. Grx1 is found mainly in the cytosol of cells and Grx2 is found in the mitochondria, the cytosol or the nucleus. These monothiols differ from the bacterial ones in that the active site may contain a single cysteine or be a Trx-Grx hybrid.

Glutaredoxins have been found to be associated with some human diseases. Glrxs protect cardiomyocytes and endothelial cells against apoptosis induced by oxidative stress, and also prevent cardiometabolic dysfunctions, which suggests their potential therapeutic effect in cardiovascular diseases. The participation of glutaredoxins in the processes taking place in the central nervous system, mainly in the protection of neurons, cell aging and neurodegenerative diseases, has also been confirmed. These compounds have also been found to be involved in the regulation of the cell cycle, drawing scientists' attention to their role in cancer development (Lee et al. 2020).

Table 1. Human glutaredoxins

Glutaredoxin	Masa [kDa]	Location	Coding gene
Grx1	12	Cytosol, mitochondrial mesothelial space, cell nucleus	GLRX
Grx2	18	Mitochondrion (Grx2a), cytosol (Grx2c), nucleus (Grx2b)	GLRX2
Grx3	37	Cytosol, the cell nucleus	GLRX3
Grx5	17	Mitochondria, cytosol	GLRX5

Although a lot of information about the properties and effects of glutaredoxins has already been discovered and described by scientists, there is still a lot that we do not know about this group of compounds. Research on the topic of glutaredoxins should focus on a greater understanding of the physiological function of these small enzymes, and more specifically in the enzymatic glutathionylation and deglutathionylation processes. Protein glutathionylation has been recognized as an important post-translational process regulating the performance of various functions by the resulting proteins. However, the mechanism by which dithiol glutaredoxins are activated and able to recognize the appropriate target proteins has not been understood so far. Moreover, most of the proteins found that undergo glutathionylation are cytosolic proteins, little is known yet about proteins in other cell compartments. Research into the role and mode of action of monothiol glutaredoxins is ongoing, but these compounds have great potential because of their key role in iron transport and homeostasis. The functions of glutaredoxins discovered so far as a probable target of drugs in several pathological processes are now a high-profile topic in the world of science, but at the same time extremely difficult to work out due to the biochemical properties of glutaredoxins and their unknown activation mechanism. Many years have passed since the first discovery made by Professor Arne Holmgren of these small enzymes in 1976 and since then many scientists have become involved in research on glutaredosins. However, there are still many unanswered questions (Matsui et al. 2020).

References

Burns M, Rizvi SHM, Tsukahara Y, Pimentel DR, Luptak I, Hamburg NM, Matsui R, Bachschmid MM. Role of Glutaredoxin-1 and Glutathionylation in Cardiovascular Diseases. *Int. J. Mol. Sci.* 2020 Sep 16;21(18):6803.

Gorelenkova Miller O, Mieyal JJ. Critical Roles of Glutaredoxin in Brain Cells-Implications for Parkinson's Disease. *Antioxid. Redox Signal.* 2019 Apr 1;30(10):1352-1368.

Jacquot JP, Zaffagnini M. Thioredoxin and Glutaredoxin Systems Antioxidants Special Issue. *Antioxidants* (Basel). 2019 Mar 18;8(3):68.

Lee J, You JH, Shin D, Roh JL. Inhibition of Glutaredoxin 5 predisposes Cisplatin-resistant Head and Neck Cancer Cells to Ferroptosis. *Theranostics.* 2020 Jun 19;10(17):7775-7786.

Matsui R, Ferran B, Oh A, Croteau D, Shao D, Han J, Pimentel DR, Bachschmid MM. Redox Regulation *via* Glutaredoxin-1 and Protein *S*-Glutathionylation. *Antioxid. Redox Signal.* 2020 Apr 1;32(10):677-700.

Chapter 18

Lactase

Bartosz Ziobro, David Aebisher and Dorota Bartusik-Aebisher*

Medical College of the University of Rzeszów, Poland

Abstract

Lactase (LPH-Lactase phlorozin hydrolase) is a protein with a molecular weight of about 160 kDa belonging to the β-galactosidase enzymes. LPH is found in the enterocyte cell membrane that forms the brush border of the small intestine. This is a characteristic feature of mammals, with the greatest amount of LPH being found in the middle of the small intestine. The gene encoding LPH is located on the long arm of the second chromosome (2q21).

Keywords: enzymes, lactase phlorozin hydrolase, biochemistry

The enzyme is a tetramer is made of 4 protein domains linked together and anchored in the cell membrane by hydrophobic chains. LPH has two types of active sites, therefore it can perform the functions of lactase as well as phlorin hydrolase (Bayless et al., 2017). The task of this enzyme is to break down lactose, also known as milk sugar due to its presence in the milk of mammals. Lactose is a disaccharide composed of D-galactose and D-glucose linked by a β-glycosidic bond. Lactase hydrolyzes the β-glycosidic bond, resulting in the formation of monosaccharides that can be absorbed from the gastrointestinal

---

[*] Corresponding Author's Email: dbartusikaebisher@ur.edu.pl.

In: The Biochemical Guide to Enzymes
Editors: David Aebisher and Dorota Bartusik-Aebisher
ISBN: 979-8-88697-410-2
© 2023 Nova Science Publishers, Inc.

tract by enterocytes and used as a source of energy by the body. Apart from lactose, LPH also hydrolyzes bonds contained in other saccharides, e.g., in cellulose, cellotriose. The enzyme, thanks to the second type of active sites, is fully phlorisine hydrolase, allowing the digestion of β-glycosides with an alkyl chain. These functions make LPH a very important enzyme for the life of mammals (Forsgård et al., 2019).

The production of lactase begins with the transcription of the gene encoding it on the second chromosome (2q21). There is a four-fold internal repeat in the gene (parts I-IV) of which the part of the polypeptide encoded by I and II is removed by post-translational processing (Leis et al., 2020). During the translation process taking place in the cytoplasm, pre-proLPH with a molecular weight of 215 dKa is produced on the ribosomes. Pre-pro LPH consists of 5 domains:

1) N-end (Pre part): made of 19 amino acids, responsible for directing the mature protein for further modification.
2) Part Pro (LPH α): this domain is not part of the mature protein (coded by parts I and II).
3) Extracellular part (LPHβ): the part of the enzyme found in the intestinal lumen (LPHβ initial is encoded by parts II-IV, and LPHβ final only by III and IV).
4) Short hydrophobic chain: enzyme domain built into the cell membrane.
5) C-end: part of the enzyme anchored in the cytoplasm.

After translation is complete, the protein is directed for further modification in the cytoplasm by the signaling sequence (N-terminus) where it is detached from the rest of the protein resulting in the formation of pro-LPH. Pro-LPH is glycosylated in the Golgi apparatus. As a result of these transformations, a glycoprotein is formed (Ségurel et al., 2017). The resulting molecule dimerizes. Mature Pro-LPH undergoes two proteolytic decays. During the first one, taking place in the cytoplasm, some LPH α is disconnected. After that, the enzyme is incorporated into the enterocyte cell membrane, where it undergoes a second proteolytic breakdown under the influence of trypsin contained therein.

As a result, LPHβ initial becomes LPHβ final). in this way, a mature LPH enzyme is created, consisting of 4 protein domains:

- *Two domains III*: located closer to the membrane, encoded by part III of the gene, performing the functions of a phlorisine hydrolase.
- *Two domains IV*: located further from the cell membrane, encoded by part IV of the gene, perform the function of lactase.

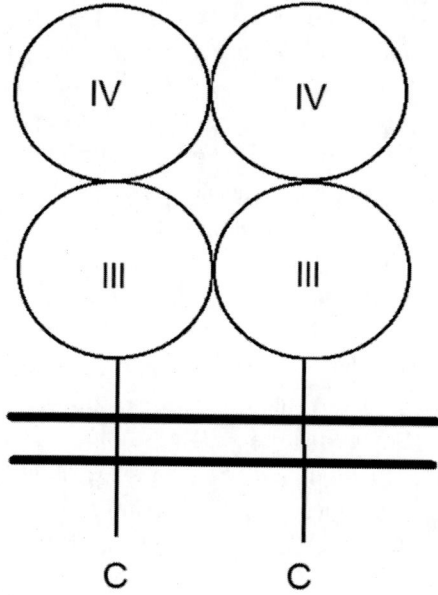

Figure 1. Diagram of the enzyme lactase in the entorocyte membrane.

Mammals produce lactase early in life, with the exception of humans, about 35% of whom do not lose the ability to produce this enzyme during their lifetime, a trait called LP (lactase persistence), which results from a mutation in the promoter of the gene encoding LPH, resulting in the production of lifelong lactase. Its incidence varies greatly between populations and ranges from 5% to even 100% (Ségurel et al., 2017). This frequency is correlated with livestock farming, especially cattle, and is highest in central and northern Europe and lowest in southern Africa and Southeast Asia. The consumption of lactose-containing products by a person without a sufficient amount of LPH causes symptoms of lactose intolerance, such as: flatulence, colic, diarrhea (Stourman et al., 2018). The cause of these symptoms is that undigested lactose is fermented by bacteria in the digestive system. The products of this fermentation, such as methane, hydrogen, carbon dioxide and organic acids, e.g., lactic acid, are the cause of the above-mentioned symptoms. Due to the reason, lactose intolerances are divided into:

- *Primary lack of lactase (alactasia)*: a rare genetic condition results in a complete lack of lactase production. Patients do not digest lactose, even in the smallest amounts, Their consumption of products with lactose may be life or health threatening.
- *Secondary lactase deficiency*: occurs much more often than the lack of the enzyme, is caused by damage to the intestinal epithelium and disappears with its regeneration. This damage may be the result of inflammation of the lining of the intestines, the use of various medications, or inflammatory bowel disease (Crohn's disease).
- *Primary lactase deficiency*: the result of the loss of lactase production with age. This is due to the lack of a mutation that would continuously produce lactase. It occurs in most of the population.

Regardless of the cause of intolerance, the symptoms are similar, and those affected should avoid lactose in larger amounts or take preparations containing lactase (Stourman et al., 2018).

Lactase, thanks to its enzymatic properties allowing it to break down lactose into monosaccharides, plays a very important role for mammals. It is even indispensable for the proper development of the representatives of this cluster. In the vast majority of mammals, lactase activity in the intestine declines with age, the exception here is humans, in which lactase activity may persist throughout life. However, this phenotype does not occur in all people, but in only ⅓ of the population. The occurrence of this feature is conditioned by mutations in the gene encoding the lactase. Lactase, due to its importance for humans, has been of great interest to researchers since the 1970s, with the first works about it being written at the beginning of the 20th century. In total, about 35,000 of them were created, of which 7,500 in the last 10 years, which indicates that lactase is still of interest to modern science.

References

Bayless T M, Brown E, Paige D M. Lactase Non-persistence and Lactose Intolerance. *Curr. Gastroenterol. Rep.* 2017;19(5):23.

Forsgård R A. Lactose digestion in humans: intestinal lactase appears to be constitutive whereas the colonic microbiome is adaptable. *Am. J. Clin. Nutr.* 2019;110(2):273-279.

Leis R, de Castro M J, de Lamas C, Picáns R, Couce M L. Effects of Prebiotic and Probiotic Supplementation on Lactase Deficiency and Lactose Intolerance: A Systematic Review of Controlled Trials. *Nutrients.* 2020;12(5):1487.

Ségurel L, Bon C. On the Evolution of Lactase Persistence in Humans. *Annu. Rev. Genomics Hum. Genet.* 2017;18:297-319.

Stourman N, Moore J. Analysis of lactase in lactose intolerance supplements. *Biochem. Mol. Biol. Educ.* 2018;46(6):652-662.

Chapter 19

Neuron Specific Enolase (NSE)

Klaudia Dynarowicz, David Aebisher and Dorota Bartusik-Aebisher[*]
Medical College of the University of Rzeszów, Rzeszów, Poland

Abstract

> Neuron Specific Enolase (NSE) is a metal-activated metalloenzyme. NSE is one of five isomers of the glycolytic enzyme enolase.
>
> **Keywords:** Neuron Specific Enolase, enzymes, biochemistry

Neuron Specific Enolase (NSE) was discovered by Lohmann and Meyerhof in 1934 (Isgrò, et al., 2015). Immediately after the discovery, it was assumed that neuron-specific enolase is the most acidic isoenzyme of the glycolytic enzyme enolase, which occurs only in central neurons. After numerous studies, it has been proven that it is also found in peripheral autonomic nerves and in endocrine cells. Two of the three subunits that make up the enolase family are found specifically in neurons. NSE is a tumour marker. NSE is also considered as a sensitive marker for evaluating the severity of craniocerebral injury, detecting small cell lung cancer and prognosis of relevant diseases. It is found in central and peripheral neurons, as well as in neuroendocrine cells, but not in glial cells. Like the 14-3-3 proteins, this enzyme can be released into the cerebrospinal fluid (CSF) when neurons are damaged. NSE is also found in trace amounts in the cytoplasm, erythrocytes, platelets and neuroendocrine

---

[*] Corresponding Author's Email: dbartusikaebisher@ur.edu.pl.

In: The Biochemical Guide to Enzymes
Editors: David Aebisher and Dorota Bartusik-Aebisher
ISBN: 979-8-88697-410-2
© 2023 Nova Science Publishers, Inc.

cells. This is due to the severity of the injury or may predict clinical outcomes in severe patients. Still, its clinical utility as biomarker is limited due to the lack of specificity as multiple factors can influence systemic NSE levels, including hemolysis, which may be prevalent in polytrauma patients.

Its main task is to convert glucose to the appropriate metabolites. In these processes, it produces both high-energy ATP compounds and the cofactor NADH. Enolase is a well-known glycolytic enolase enzyme which catalyzes 2-phosphoglycerate to phosphoenolpyruvate. The group of enolases is one of the most ubiquitous and most abundantly expressed proteins in cells. NSE is a sensitive indicator for assessing the severity of nerve cell damage and prognosis.

The active forms of enolase are dimers. There are mainly three enolase isoenzymes in the organisms of vertebrates: α, β, and γ-enolase (Kaiser, et al., 1989). Enolases are located mainly in the cytoplasm of the cell and exert a catalytic effect. Another place of its presence is: the cell nucleus, the cell membrane and the extracellular space (Xu, et al., 2019). Cell surface enolase expression is often detected on activated macrophages, microglia, neurons and astrocytes, leading to the activation of various cellular events and the production of pro-inflammatory cytokines (Haque, et al., 2016). Cell surface enolase induces surface bound proteolysis, which promotes the migration of phagocytic cells to the site of inflammation. NSE, as a neuronal enolase of a glycolytic enzyme, occurs almost exclusively in neurons and cells of neuroendocrine origin (Mjønes, et al., 2017).

It influences neurotrophic activity and regulates the growth, differentiation, survival and even regeneration of neurons. Under normal physiological conditions, the NSE neuron is not secreted into the extracellular space but may penetrate it or be regulated in response to damaged neuronal tissue. NSE is a marker of many pathological conditions. Its presence and content is assessed in such diseases as: neuroendocrine neoplasms, brain damage resulting from cardiac arrest, ischemic stroke. NSE not only informs about neuronal changes in the body, but also in the case of diagnosing malignant pleural effusion. It is generally accepted that NSE is a useful biomarker of Creutzfeldt-Jakob disease, hypoxic encephalopathy, epilepsy, and brain trauma.

Elevated levels also occur in other clinical conditions. Table 1 presents clinical conditions and neoplastic diseases with increased levels of NSE.

High concentration of NSE occurs in glioblastomas, it is sometimes elevated after head injuries and septic shock. In small cell lung cancer, the concentration of NSE is increased in 90% of cases of systemic disease and in 60% of limited form cases.

The half-life of NSE is approximately 24h. The level of NSE depends on the weight of the tumor, the number and type of metastasis, and the body's response to treatment.

Table 1. Clinical conditions in which NSE is elevated

Clinical conditions in which NSE is elevated
Melanoma
Seminoma
Renal cell carcinoma
Merkel cell tumour
Carcinoid tumours
Dysgerminomas
Immature teratomas
Malignant phaeochromocytomas
Ischaemic stroke
Intracerebral haemorrhage
Inflammatory brain diseases
Creutzfeldt-Jakob disease

Neuron-specific enolase is not only a specific molecular marker for mature nerve cells but is closely correlated to the differentiated state.

NSE is an old marker that is sensitive but not very specific. From the research, it can be concluded that NSE may be a candidate for a diagnostic / prognostic biomarker of neurological inflammation in COVID-19, especially in patients with neurological symptoms.

Due to the characteristics of NSE, it is a very helpful tool that enables the characterization of many clinical and disease states.

References

Haque, A., Ray, S. K., Cox, A., Banik, N. L. Neuron specific enolase: a promising therapeutic target in acute spinal cord injury. *Metab. Brain Dis.* 2016, 31, 487-495.

Isgrò, M. A., Bottoni, P., Scatena, R. Neuron-Specific Enolase as a biomarker: biochemical and clinical aspects. *Adv. Exp. Med. Biol.* 2015, 867, 125-143.

Kaiser, E., Kuzmits, R., Pregant, P., Burghuber, O., Worofka, W. Clinical biochemistry of neuron specific enolase. *Clin. Chim. Acta.* 1989, 183, 13-31.

Mjønes, P., Sagatun, L., Nordrum, I. S., Waldum, H. L. Neuron-Specific Enolase as an immunohistochemical marker is better than its reputation. *J. Histochem. Cytochem.* 2017, 65, 687-703.

Xu, C. M., Luo, Y. L., Li, S., Li, Z. X., Jiang, L., Zhang, G. X., Owusu, L., Chen, H. L. Multifunctional neuron-specific enolase: its role in lung diseases. *Biosci. Rep.* 2019, 39.

Chapter 20

Hyaluronidase

Anna Zdziebło, David Aebisher and Dorota Bartusik-Aebisher[*]

Medical College of the University of Rzeszów, Poland

Abstract

Hyaluronidase is a protein with enzymatic properties that catalyzes the degradation of hyaluronic acid. Thanks to its ability to bind water, it has strong moisturizing properties.

Keywords: hyaluronidase, enzymes, biochemistry, enzymes

At the outset, it is worth noting how common in the human body is hyaluronic acid, which is the most abundant glycosaminoglycan. It is a polymer with alternating units of D-glucuronic acid and N-acetyl-D-glucosamine (Helm et al., 2019). It has the same structure in all living organisms. It is an essential component of the basal membrane of cells and the intercellular matrix of all tissues. It is present in large amounts in the synovial fluid where it reduces friction and nourishes the cartilage. It is also used in aesthetic medicine, where it is used as a filler to reduce, for example, wrinkles or furrows, as well as to enlarge the lips or correct the nose; in dermatology for the treatment of burns and various types of scars. Its ubiquitous use highlights the importance of the existence of hyaluronidase, endoglycosidase, an enzyme that depolymerizes hyaluronic acid. It cleaves the bonds (1-4) between N-acetylglucosamine and

---

[*] Corresponding Author's Email: dbartusikaebisher@ur.edu.pl.

In: The Biochemical Guide to Enzymes
Editors: David Aebisher and Dorota Bartusik-Aebisher
ISBN: 979-8-88697-410-2
© 2023 Nova Science Publishers, Inc.

glucuronide (Helm et al., 2019). Thanks to these properties, hyaluronidase increases the permeability of connective tissue, which results in a faster and more extensive spread of various substances in the human body. This enzyme also allows the dissolution of hyaluronic acid in the patient after an unsuccessful aesthetic procedure. Currently, only five types of hyaluronidase have been discovered and described: HYAL1, HYAL2, HYAL3, HYAL4 and HYAL 5, which are most effective at acidic pH. Mutations in genes encoding enzymes or their deficiency are often the cause of serious diseases (Locke et al., 2019).

The action of hyaluronidase was first described in 1936 by Duran-Reynals from Barcelona, who called the enzyme a "spreading factor" because of its ability to break down the structure of hyaluronic acid and increase tissue permeability (Locke et al., 2019). Then in 1971. German scientist Karl Meyer, conducting research on connective tissue, classified and divided hyaluronidases into three groups. Each of them differed in the chemical composition and the resulting end products of the reaction, i.e., to mammals, leeches and bacteria hyaluronidase. In humans, 5 genes encoding hyaluronidase (HYAL1-5) have been identified. It is believed that there is a 6 variant - HYAL6, but for now it is referred to as a pseudogene (Schinzel et al., 2019). The genes encoding HYAL1-3 are clustered on chromosome 3, while HYAL4 and HYAL5 are clustered on chromosome 7. Of all, HYAL1 and HYAL2 are the major hyaluronidases responsible for the catabolism of hyaluronic acid in connective tissue. While HYAL3 to 5 are inactive and are not likely to be involved in enzymatic acid hydrolysis. Human HYAL3 is expressed in several tissues, including the brain, but its enzymatic activity and biological role are not well defined. Human HYAL4 has been reported to be specific for chondroitin and chondroitin sulfate substrates present in abundance in cartilage tissue (Schinzel et al., 2019). The described sequences of known human hyaluronidases (Hyal-1-5) are relatively uniform in amino acid length and range from the shortest, HYAL1 with 435 amino acids, to the longest, HYAL4 with 510. All human hyaluronidases analyzed so far are homologous to each other. The structure of hyaluronidase includes N-terminal, linker and C-terminal domains. The active site is in a slot in the N-terminal domain. Two molecules of hyaluronic acid are bound at the active site, making contact with some enzyme residues and repeatedly contacting the water molecules (Weber et al., 2019).

Figure 1. Structure of DNA.

As mentioned before, hyaluronidase cleaves beta-1-4 glycosidic bonds of hyaluronan glycosamin. Injection of hyaluronidase with other fluids, drugs, or radiopaque agents improve the ability of these compounds to penetrate the extracellular space more easily (Weber et al., 2019). Hyaluronidase has been the subject of interest of a large number of scientists for years, its ubiquitous presence prompts them to research its activity and application. Today, you can find many publications, scientific articles and clinical research results on it.

Much research is available on hyaluronidase produced by bacteria. Some of them, such as Staphylococcus aureus, Streptococcus pyogenes and Clostridium perfringens, produce hyaluronidase to extract carbon from hyaluronic acid. Scientists also believe that Streptococcus and Staphylococcus bacteria use hyaluronidase as a virulence factor to destroy the polysaccharide that binds animal cells, making it easier for them to spread through the host's tissues. The physiological role of hyaluronidases is also played in the fertilization of mammals. Hyaluronidase is released by the acrosome from the sperm after reaching the oocyte. The enzyme then digests the corona granule cells, which in mammals contain a high concentration of hyaluronan. Thanks to this, sperm are able to bind with the transparent sheath, which is the next step to fertilization (Wang et al., 2017).

Leukocytes (white blood cells) also produce hyaluronidase. Thanks to it, they move more easily through the connective tissue and can get to infected places and fight the pathogen. In medicine, hyaluronidase has many important functions. It is widely used in aesthetic medicine and allows the dissolution of hyaluronic acid, which results in the elimination of undesirable effects after the introduction of the filler. It is also commonly used in surgery to better spread anesthetics and to eliminate tumors of connective tissue (Wang et al., 2017).

Hyaluronidase deficiency (mucopolysaccharidosis type IX) is a pathological condition caused by a mutation in the HYAL1 gene. The disease is very rare because it has a frequency of less than 1: 1,000,000. Symptoms of this condition may include:

- Multiple soft tissue tumors
- Transient episodes of genealized skin swelling
- Low rise

Summary

The multi-molecular protein hyaluronidase has found a number of applications thanks to its catalytic properties to accelerate the decomposition of hyaluronic acid. Its presence is essential for the proper functioning of many organisms, not only humans, and its ubiquitous availability enables its use in many fields of medicine. Further clinical and experimental studies may provide better information on the new and diverse therapeutic uses of hyaluronidase in clinical practice in the future. The US Food and Drug Administration in the 1960s officially approved hyaluronidase as a substance used for the following indications: subcutaneous fluid infusion, as a means to accelerate the absorption and dispersion of drugs in the subcutaneous tissue or to control extravasation, and as an additive to support the absorption of contrast agents in angiography urinary tract.

In addition, it has been approved and used to increase the absorption of subcutaneous hematomas by agencies in Europe. Hyaluronidase has been used for years to treat the side effects of improper placement of hyaluronic acid filler in patients. The use of preparations containing hyaluronidases is generally well tolerated by patients and side effects are very rare. However, a few cases of undesirable effects associated with the use of hyaluronidase have been reported, such as transient injection pruritus or allergic reactions, local or very rarely generalized. There is no doubt that hyaluronidase is a key protein in the course of many physiological processes, such as, for example, fertilization, and also significantly facilitates the work of a doctor and improves patient comfort, by facilitating the spread of drugs, e.g., painkillers, and faster absorption of hematomas. It is a relatively safe preparation due to its natural production and common occurrence in the human body. Its advantage is also quick decomposition and deactivation in the body, thanks to which it works only in a specific place and time.

References

Helm Ii S, Racz G. Hyaluronidase in Neuroplasty: A Review. *Pain Physician.* 2019;22(6): 555-560.

Locke K W, Maneval D C, LaBarre M J. ENHANZE® drug delivery technology: a novel approach to subcutaneous administration using recombinant human hyaluronidase PH20. *Drug Deliv.* 2019;26(1):98-106.

Schinzel R T, Higuchi-Sanabria R, Shalem O, Moehle E A, Webster B M, Joe L, Bar-Ziv R, Frankino P A, Durieux J, Pender C, Kelet N, Kumar S S, Savalia N, Chi H, Simic M, Nguyen N T, Dillin A. The Hyaluronidase, TMEM2, Promotes ER Homeostasis and Longevity Independent of the UPRER. *Cell.* 2019;179(6):1306-1318.e18.

Wang W, Wang J, Li F. Hyaluronidase and Chondroitinase. *Adv. Exp. Med. Biol.* 2017;925:75-87.

Weber G C, Buhren B A, Schrumpf H, Wohlrab J, Gerber P A. Clinical Applications of Hyaluronidase. *Adv. Exp. Med. Biol.* 2019;1148:255-277.

Chapter 21

Beta-Galactosidase

Mateusz Warzocha, David Aebisher and Dorota Bartusik-Aebisher[*]
Medical College of the University of Rzeszów, Rzeszów, Poland

Abstract

Beta-galactosidase is an enzyme that commonly occurs in the human organism. It is taking part in the digestion of lactose, which may sometimes cause disorders of this process. New research points to the involvement of beta-gal in the cell aging processes. There are many methods of beta-gal activity detection, one of which is fluorescent. As the biomarker, it has been used for ovarian carcinoma cells detection. It also takes part in colorectal tumor imaging. Therefore, in the future it may apply for diagnosis. The application of this enzyme has also been found in the food industry, especially to crack bonding in complex sugars.

Keywords: beta-galactosidase, emzymes, biochemistry

One of the older information available about beta-galactosidase is a book published in 1945 that the enzyme obtained from sweet almonds breaks down lactose (4-beta-D-galactosido-D-glucose) much more intensively than neolactose (4-beta-D-galactoside-D-altrose). Obviously, this topic has appeared with increasing intensity over the years, which proves that it is probably widely used in various fields. Beta-galactosidase belongs to the

---

[*] Corresponding Author's Email: dbartusikaebisher@ur.edu.pl.

In: The Biochemical Guide to Enzymes
Editors: David Aebisher and Dorota Bartusik-Aebisher
ISBN: 979-8-88697-410-2
© 2023 Nova Science Publishers, Inc.

hydrolases that break down lactose most efficiently, therefore its common name is lactase. Its main feature is the cleavage of the beta-D-galactoside bond (β-D-galactoside), it also has transgalactosylation activity during the synthesis of galacto-oligosaccharides. It occurs in plants, fungi, bacteria, as well as animals and humans. It is classified in the family of exoglycosidases because it has the ability to eliminate galactose residues from various substances present in the human body, such as glycoproteins, gangliosides and sphingolipids (Damin et al., 2021).

The genetic information encoding the lactase is located on the long arm of chromosome 2 (2q21 region) of the human genome. Its expression is regulated by the promoter region upstream of the gene.

Breast milk contains 7.5 g / ml of lactose, its breakdown requires the production of lactase, which takes place in the brush border, in the jejunum. A newborn baby breaks down 60-70g of lactose daily, so beta-gal is essential for the proper functioning of the digestive system and baby's development. With age, the amount of lactase disappears, causing various ailments. However, in some adults there is a single nucleotide polymorphism (SNP) in a regulatory element, which leads to the digestion of lactose during their adult life. This mutation is common in people of North European, West African or Middle Eastern descent (Jannone et al., 2020).

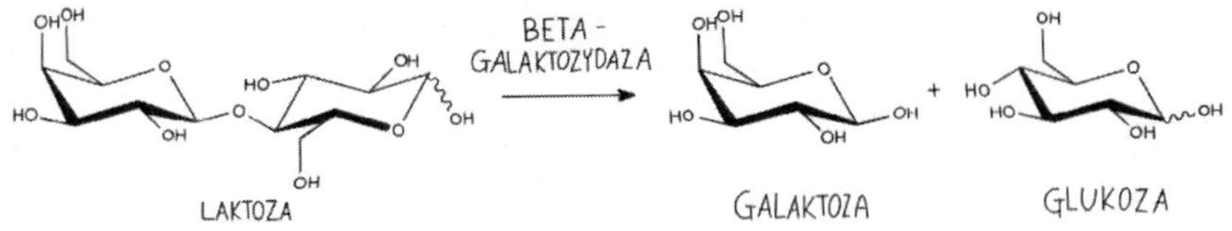

Figure 1. Diagram showing the enzymatic degradation of lactose.

Estimates of the prevalence of reduced production or absence of the lactase enzyme are sequentially 2–5% in Northern Europe (Scandinavia, Germany, UK), 17% in Finland and Northern France, around 50% in South America and Africa, and between 90 and 100% in Southeast Asia. However, in North America, adult lactase production is dependent on ethnicity.

Lack of lactase leads to an increase in lactose in the large intestine, which in turn causes an influx of water into the lumen of the canal, thus causing diarrhea. Diluted food content with undissolved sugar is an excellent breeding ground for bacteria that produce hydrogen, methane, carbon dioxide and short-chain fatty acids, which causes flatulence and abdominal pain. Beta-galactosidase can be detected in vivo and in vitro using many techniques and research methods, these are bioluminescence, magnetic resonance (MR), chemiluminescence, positron emission tomography (PET), single photo-emission computed tomography, colorimetric and fluorogenic methods. Among the above methods, fluorescence deserves attention, it is characterized by high sensitivity, low cost, easy application and the possibility of assessing the enzyme in living cells. Thanks to it, it is possible to observe the dynamics, physiology and pathology of a given enzyme in individual biological systems (Lozano-Torres et al., 2021).

Detection of beta-gal in human cells is possible thanks to a fluorescent near infrared (NIR) probe, disruption of the groups and cleavage of a new structure with fluorescent properties, thanks to Beta-galactosidase. This technique is used to evaluate cells, including the identification of irregular overexpression of certain genes in cancer cells.

Bio-imaging of various enzymes, including beta-gal, can be used for in-depth diagnosis of various diseases. Currently, there are reports that beta-gal serves as a biomarker of ovarian cancer cells, in order to identify abnormal cells, a compound is used, which is hydrolyzed by the said enzyme, leading to the formation of a compound with fluorescence properties. This method can be successfully used to detect an ongoing disease process. Another application of this technique is the in situ and in vivo detection of β-gal activity in human colon cancer in a mouse model. In addition, three-dimensional, real-time visualization of beta-galactosidase activity can be used. This unprecedented technique could be used to evaluate the uptake of enzyme activity for the diagnosis of colorectal cancer in humans (Sharma et al., 2021).

The lysosomal activity of beta-galactosidase belongs to the biomarkers of aging, for this purpose, a compound that kills aging was synthesized, "senescence-specific killing compound 1" (SSK1), which, under the influence of the lysosomal enzyme beta-gal was dissolved into a cytotoxic gemcitabine. The SSK1 substrate, through its enzymatic degradation and its products, lyses aging cells, but it does not act on non-aging cells. The study was carried out on mouse and human cells, however, the aging of the organism is a complex physiological and psychological process in which many systems are involved. Currently, the use of beta-gal for the production of low-lactose ice cream with probiotic bacteria containing *Bifidobacterium animalis* ssp. Lactis as functional food is considered. Despite the low effect of the enzyme on the growth of bacteria, beta-galactosidase breaks down lactose contained in milk, so it can be used as a substrate for the production of ice cream, devoid of this compound. The key issue is the more intense sweet taste of low-lactose ice cream compared to lactose ice cream, which will limit the addition of sucrose during production and, consequently, reduce the consumption of carbohydrates (Uchil et al., 2017).

Beta-galactosidase is a protein that has dominated the scientific community, spanning science from industry to medicine. The first mentions of it appeared in the first half of the 20th century and concerned its influence on sugars. Typing the term "beta-galactosidase" in the PubMed.gov search engine brings up 14,488 results which proves that you are constantly learning about beta-gal, and more interestingly, getting new, exciting and amazing results.

Along with the development of equipment, new measurement techniques, chemical compounds and innovative technologies, there has been an explosion of interest in researching new possibilities of this enzyme. Thus, there are many publications on the use of beta-galactosidase as a biomarker of cancer diseases, an enzyme used at various stages of the food industry or in the diagnosis of genetic diseases, e.g., lactose intolerance.

One popular technique for detecting in vivo and in vitro activity of beta-gal is fluorescence. Thanks to it, it is possible to detect metabolically changed cells that have mutations. This makes it possible to detect colorectal cancer, ovarian cancer or other diseases not included here.

The influence of beta-gal together with other substances on the process of lysis of senescent cells has been proven. This discovery will contribute to the development of substances that influence the aging process of cells, but the complexity of the process raises some scientists' concern.

References

Damin B I S, Kovalski F C, Fischer J, Piccin J S, Dettmer A. Challenges and perspectives of the β-galactosidase enzyme. *Appl. Microbiol. Biotechnol.* 2021;105(13): 5281-5298.

Jannone G, Rozzi M, Najimi M, Decottignies A, Sokal E M. An Optimized Protocol for Histochemical Detection of Senescence-associated Beta-galactosidase Activity in Cryopreserved Liver Tissue. *J. Histochem. Cytochem.* 2020; 68(4): 269-278.

Lozano-Torres B, Blandez J F, Sancenón F, Martínez-Máñez R. Chromo-fluorogenic probes for β-galactosidase detection. *Anal. Bioanal. Chem.* 2021; 413(9): 2361-2388.

Sharma S K, Poudel Sharma S, Leblanc R M. Methods of detection of β-galactosidase enzyme in living cells. *Enzyme Microb. Technol.* 2021; 150: 109885.

Uchil P D, Nagarajan A, Kumar P. β-Galactosidase. *Cold Spring Harb Protoc.* 2017; 2017(10): pdb.top096198.

Chapter 22

Photodynamic Therapy

Adamo Federica, David Aebisher and Dorota Bartusik-Aebisher*
Medical College of the University of Rzeszów, Poland

Abstract

Photodynamic therapy (or PDT) undoubtedly represents one of the most innovative, modern and technologically advanced dermatological therapies currently available. Initially introduced for the treatment of pre-cancerous lesions and for skin tumors deriving from the epidermis (and therefore not melanocytic), photodynamic therapy has been used to treat an increasingly broad spectrum of skin conditions, both pathological and aesthetic (the so-called photorejuvenation). Based on the use of a sensitizing agent, PDT uses a particular light source to obtain its therapeutic and cosmetic effects. From light, therefore, the cause of these types of injuries and skin aging, science has developed a technique that, through light itself, brings health and beauty to the skin.

Keyword: photodynamic therapy, mechanism, biochemistry

Photodynamic Therapy is a medical procedure based on the dermatological field with the use of a cream containing a photosensitizing agent applied to the interested area. This photosensitizing agent has the ability to penetrate the skin and selectively accumulate in the diseased cells (whether they are pre-tumor, frankly tumor or "aged" by continuous exposure to the sun) (Dobson et al.

---

* Corresponding Author's Email: dbartusikaebisher@ur.edu.pl.

In: The Biochemical Guide to Enzymes
Editors: David Aebisher and Dorota Bartusik-Aebisher
ISBN: 979-8-88697-410-2
© 2023 Nova Science Publishers, Inc.

2018). This accumulation of substance is able to make these cells produce a phototoxic substance, which is activated when the area is illuminated by a particular light source. This interaction between light and phototoxic substance produces very reactive oxygen molecules inside the cell, capable of destroying cells from the inside. Once the process is complete, the immune system inherent in our skin will cleanse the field of the residues of the destroyed cells and will also be more effective in eliminating any surviving diseased cells (Dobson et al. 2018).

From a technical point of view, the main photosensitizing agent used in skin PDT is a cream containing Amino-Levulinic Acid, in its methylated form (hence the name of Methyl-aminolevulinate or MAL). This compound is normally present in our body as it is part of the biosynthetic pathway of heme, the particular substance part of hemoglobin and capable of binding oxygen in our red blood cells. MAL has the ability to selectively accumulate in diseased cells (i.e., those with a higher metabolism) and, once present in these cells, the accumulation of this substance short-circuits the biosynthetic pathway of heme, causing the accumulation of a metabolite of waste called Proto-porphyrin IX. This substance is the real responsible for the PDT mechanism of action. In fact, when illuminated by a source of red light, it has the ability to absorb energy and enter a so-called "excited" state; to return to its state of rest, it must yield the accumulated energy to the surrounding molecules, and among these the most receptive to collecting excess energy are the oxygen molecules. When oxygen absorbs external energy, it itself enters a state of "excitation" forming the so-called ROS or reactive oxygen species. These ROS are highly toxic to the cell because they destroy everything they encounter, such as proteins, membranes, etc. Once this reaction has taken place, the cell undergoes degeneration, which can be controlled (apoptosis phenomenon) or uncontrolled (necrosis phenomenon). In both cases, the final result of photodynamic therapy is the selective destruction of the cells that have incorporated the MAL, which, as mentioned, are generally the very sick ones (Kwiatkowski et al. 2018). At present, the only substance that has obtained official registration in order to be used in photodynamic therapy and which has obtained the largest number of supporting scientific studies is precisely Methylaminolevulinate, there are also other photosensitizing substances in use for skin PDT.

The dermatological applications of photodynamic therapy are divided into official indications (i.e., those for which this therapy has been validated by international control bodies) and so-called "off-label" indications, i.e., unofficial but based on the study of experts in the sector and validated by publications international scientific studies (however not sufficient to have an official indication). The skin diseases in which the official indication for the use of photodynamic therapy has been given are:

- Actinic keratosis of slight thickness or non-hyperkeratotic and non-pigmented of the face and scalp;
- Superficial and/or nodular basal cell carcinoma (less than 2 mm thick) for which other available therapies are not indicated, due to possible treatment-associated morbidity and poor cosmetic outcome, such as lesions on the central part of the face or ears, lesions on severely sun-damaged skin, extensive or recurrent lesions;
- Squamous cell carcinoma in situ (Bowen's disease) where surgical removal is considered a less appropriate alternative.

In case of diagnostic doubt, the execution of an incisional biopsy of the lesion by punch with histological verification may be indicated (Plotino et al. 2019).

The skin conditions in which "off-label" photodynamic therapy is used but with the greatest scientific evidence of efficacy are:

- Photodamage of face, hands, neckline;
- Treatment of the so-called "cancerization field."
    - This phenomenon indicates the presence in the skin subjected to intense and prolonged sun exposure of areas of skin predisposed to the formation of pre-cancerous or frankly cancerous lesions, however not directly visible on the skin or detectable with the common diagnostic methods available. This field generally surrounds the presence of pre-cancerous or cancerous already clinically detectable, which can relapse from it to give rise to new lesions;
- Acne;
- Viral warts;
- Genital warts or warts;
- Cutaneous leishmaniasis;
- Psoriasis;
- Cutaneous lymphomas;
- Other forms of non-melanocytic skin cancers not included in the official indications.

For such conditions, treatment with photodynamic therapy must be recommended by a dermatologist experienced in the use of this therapy, to assess whether the conditions exist to obtain a clinically significant result and evaluate the costs/benefits of PDT and other therapeutic alternatives (Plotino et al. 2019).

Once the pre-treatment visit has been performed, where the actual presence of skin lesions compatible with pathologies that can benefit from the use of PDT is established, we will move on to the actual photodynamic therapy session. Generally, the two visits are spaced by a few days, in order to give the patient the opportunity to carefully read the information prospectus and sign the informed consent, possibly also evaluating with his family doctor the cost/benefit ratio and possible side effects of photodynamic therapy before undergoing the session. The patient can discuss these risks, benefits and costs with their trusted dermatologist during the preliminary visit and in any case at any time.

On the day of treatment, the patient will initially undergo the application phase of the photosensitizing cream (containing aminolevulinic acid). In this situation, it is important that female patients do not cover the affected area to be treated and at least a couple of centimeters of surrounding skin with make-up. This phase can be preceded by the curettage of very keratotic lesions, (generally performed with a curette or scalpel blade) in order to detach the thicker crusts, in order to facilitate the penetration of the cream into the skin. In the case of combined photorejuvenation techniques (such as photopeeling), chemical peels (with glycolic acid, pyruvic or other) may also be applied, which are subsequently neutralized. The application of the cream containing aminolevunic acid is carried out directly on the site of the lesion and in the 5-10 mm of surrounding healthy skin. Subsequently, an occlusive dressing is applied that prevents the passage of light. This darkening of the already medicated lesion to be treated is of fundamental importance since if the light reached the cream before it penetrated the skin, the product would be activated in an unnecessary location and therefore the therapy would be ineffective. Even sunlight, in fact, is able to produce a certain photodynamic effect, however less than that obtained with irradiation with a red-light source (Plotino et al. 2021).

Once the dressing is complete, the patient must wait about 3 hours (the waiting time may vary depending on the pathology to be treated and on the judgment of the dermatologist). During this period and depending on the area to be treated, the patient may remain in a closed and darkened place or carry out their normal work activities, being warned not to expose themselves to temperatures below 15 °C or that the dressing does not come off. During this period, physical activity and smoking are prohibited. After the incubation period, the dressing will be removed as well as any excess residual cream. The area to be treated will then be placed immediately afterwards under a red LED light source (37 J/cm^2 for narrow band LED lamps with a wavelength of 630 nm) for about 8-10 minutes. Before starting the lamp, the patient and the doctor or nursing staff present will wear special protective goggles. At the end of the irradiation time, an antibiotic cream will be applied as a precaution and the patient can return to their activities. The patient must not expose himself to the sun or to artificial light sources in the 48 hours following the photodynamic therapy; in case of local side effects (see below), the period of abstaining from the sun or artificial light could be longer (generally up to a maximum of 15 days) to allow the skin to return to its complete normalcy and avoid possible hyperchromic post-inflammatory outcomes (Zhang et al. 2018).

Are there any contraindications for carrying out photodynamic therapy?

In general, PDT can be performed on virtually all patients, since the photosensitizing drug has very limited systemic absorption and side effects are usually small.

However, there are conditions that contraindicate the execution of photodynamic therapy:

- Patients allergic to peanuts (due to an excipient of which the photosensitizing cream is made);
- Patients with porphyria;
- Patients with xeroderma pigmentosum;
- Pregnant or lactating female patients;
- Children under the age of 8;
- Use of creams containing retinoids in the previous month;
- Ascertained photosensitivity to therapeutic wavelengths (600-700 nm).

Patients taking anticoagulant therapies should be careful in the curettage phase of the lesions, which can cause some degree of bleeding; the trusted dermatologist will decide on the basis of the clinical situation whether it is necessary to suspend the anticoagulant therapy for a few days or to replace it with another therapy, in agreement with the patient's family doctor.

It is also a good idea to inform your trusted dermatologist of the drugs taken, even if there are no known interactions of photodynamic therapy with common drug therapies.

One or more of the following side effects may occur during or after the photodynamic therapy session. These phenomena are generally of modest entity, and some are indicative of the good functioning of the therapy.

During the PDT session, the most common phenomenon is the onset of a burning sensation accompanied by a tenderness of the irradiated area. This phenomenon is due to the occurrence of the phototoxic reaction and the selective destruction of diseased cells; depending on the intensity of the phototoxic reaction, the pain sensation could also become very intense. In such situations, the doctor or the staff present may intervene by cooling the area through a jet of cold air or thermal water; in cases of more severe pain, treatment can be suspended and then resumed after a few minutes. This dose splitting does not affect the final result of the therapy (Zhang et al. 2018).

In the days following the therapy it is possible (but not necessarily) to witness the appearance of the following phenomena:

- In the 24 hours following the treatment, itching, burning or even pain may persist; in some cases, the pain may increase in intensity, and then gradually decrease. This situation can generally last up to 5-7 days. This phenomenon can be countered with the use of usual painkillers (eg paracetamol, ibufrofen, salicylic acid).
- In the 2-5 days following the treatment, peeling and crusting phenomena may appear; sometimes (but less frequently) small fluid-filled blisters and pustules may also appear. This situation can typically last up to 15 days. This phenomenon can be counteracted and prevented with the antibiotic cream prescribed after the treatment.
- In the following 24 hours and after a few days, swelling with edema may appear in the treated area. This situation can last up to 15 days. This phenomenon is due to the inflammation caused by the phototoxic reaction. If the treatment was done on the forehead, the swelling may extend to the eye area. This phenomenon can be counteracted with the use of cold compressed thermal water in the affected areas.

Infrequent but possible side effects of photodynamic therapy can be the following:

- Appearance of hyperchromic discolorations (with the appearance of dark spots on the phototoxic reaction zone. These phenomena generally lessen after a few months) and hypochromic (with the appearance of lighter spots than the surrounding skin. Also, these phenomena generally fade after few months). This phenomenon may be due to the skin repair processes after the phototoxic reaction and can be counteracted by the regular use of protective sun creams with a protection factor of 50+ for about 3 months.
- Appearance of allergic contact dermatitis or urticarial reactions due to hypersensitivity to any of the components of the photosensitizing cream; in general this phenomenon is evident in those patients who have carried out several sessions very close in time.

In addition to the local side effects mentioned above, the following general symptoms are also possible, such as: anxiety, headache, nausea, fatigue, flu-like syndrome.

If these phenomena seem excessive or not well controlled, it will be necessary to contact your trusted dermatologist promptly, in order to be able to establish adequate therapeutic measures.

It should also be noted that the appearance of such phenomena is often correlated to the depth of the treated lesion and to the same nature. Cancerous or pre-cancerous skin diseases can more frequently favor the onset of such side effects, since in such conditions the phototoxic reaction is always of greater intensity.

Photodynamic Therapy is generally effective and well tolerated by all subjects. It is contraindicated in patients with photosensitivity, porphyria or xeroderma pigmento sum and in those who have received retinoic acid treatment in the previous month.

Patients who have to perform the treatment must avoid the use of cosmetics, make-up, lotions, creams and deodorants on the affected part starting three days before the treatment. On patient is then applied the cream with aminolevunic acid on the interested area, which is bandaged. Once this operation is completed, the patient must wait about three hours to allow the cream to penetrate the skin and interact with the damaged cells. During this period the patient should not exercise and should not smoke. Exposure to red light lasts ten minutes.

References

Dobson J, de Queiroz GF, Golding JP. Photodynamic therapy and diagnosis: Principles and comparative aspects. *Vet J.* 2018; 233: 8-18.

Kwiatkowski S, Knap B, Przystupski D, Saczko J, Kędzierska E, Knap-Czop K, Kotlińska J, Michel O, Kotowski K, Kulbacka J. Photodynamic therapy - mechanisms, photosensitizers and combinations. *Biomed Pharmacother.* 2018; 106: 1098-1107.

Plotino G, Grande NM, Mercade M. Photodynamic therapy in endodontics. *Int Endod J.* 2019; 52(6): 760-774.

Warrier A, Mazumder N, Prabhu S, Satyamoorthy K, Murali TS. Photodynamic therapy to control microbial biofilms. *Photodiagnosis Photodyn Ther.* 2021; 33: 102090.

Zhang Q, Li L. Photodynamic combinational therapy in cancer treatment. *J BUON.* 2018; 23(3): 561-567.

Chapter 23

Methods of Diagnostics of Ulcerative Colitis

**Agnieszka Przygórzewska,
Iga Serafin, Kacper Rogóż,
Paweł Woźnicki, David Aebisher
and Dorota Bartusik-Aebisher**[*]
Medical College of the University of Rzeszów, Poland

Abstract

Ulcerative colitis (UC) is a chronic inflammatory disease of the colon, belonging to inflammatory bowel diseases. Its etiology is unknown. The first reports of a disease with ulceration of the mucosa of the large intestine, which was also the cause of death of patients, come from 1793. Instead, the term ulcerative colitis was used by the British physician and researcher Samuel Wilks in 1859. He described the case of a 42-year-old woman who died after several months of diarrhea with a fever that was difficult to reduce. In a postmortem examination, he showed that the cause of the symptoms was ulcerative colitis and the terminal ileum.

Keywords: enzyme, disease, treatment, biochemistry

About a hundred years later, this case was classified as Crohn's disease. In 1902 RF Weir described appendectomy in a patient with ulcerative colitis. A breakthrough in the diagnosis of UC was the discovery of an electronically illuminated proctosigmoidoscope. In the 20th century, the number of reported

---

[*] Corresponding Author's Email: dbartusikaebisher@ur.edu.pl.

In: The Biochemical Guide to Enzymes
Editors: David Aebisher and Dorota Bartusik-Aebisher
ISBN: 979-8-88697-410-2
© 2023 Nova Science Publishers, Inc.

cases of this disease continued to increase. Nowadays, UC is most common in North America and Europe, which suggests that the disease is related to the Western lifestyle. Annually, from 156 to 291 cases of UC are registered per 100,000 people. Every year, between 9-20/100,000 new patients are added, mainly in patients between the ages of 15 and 30 and, to a lesser extent, between the ages of 50 and 70. Most studies do not indicate the influence of the patient's sex on the incidence of UC. Ulcerative colitis is clinically diagnosed on the basis of endoscopic examination, biopsy and negative stool examination for infectious causes. Since colon infection may have clinical symptoms indistinguishable from UC, the initial assessment should also include microbiological tests of bacterial infection or parasitic infestation (Feuerstein et al. 2020).

The aim of the study was to review the literature on the methods of diagnosing ulcerative colitis.

Laboratory tests, including blood tests, are an essential part of the diagnosis of ulcerative colitis. Blood tests should take into account the measurement of inflammatory factors, i.e., the number and smear of leukocytes, the number of thrombocytes, CRP and ESR. The classic parameters of inflammation (leukocyte count, ESR, and CRP) are generally not elevated unless the inflammatory activity of ulcerative colitis is very intense. It follows that elevated parameters of inflammation imply a severe course of the disease. In this case, hypoalbuminaemia may also occur, which may herald the need for a colectomy later. Patients may develop leukocytosis and thrombocytosis, but this is not always the case. In addition, a complete blood count should be performed, the concentration of ferritin and transferrin saturation should be determined, because the occurrence of anemia due to iron deficiency is one of the factors indicating a probable inflammatory bowel disease (Flynn et al. 2019).

Laboratory tests, including stool examination, are part of the diagnosis of ulcerative colitis. The concentration of calprotectin (norm <50 μg/g) or lactoferrin should be determined there. However, the content of these proteins is not a clear indicator of the occurrence of ulcerative colitis - a low level of calprotectin only makes it possible to exclude inflammatory bowel disease (probability less than 1%). On the other hand, the amount of calprotectin lower than 150–200 μg per gram of stool indicates remission of the disease. In patients with suspected ulcerative colitis, stool culture for Clostridium difficile should be performed to exclude bacterial infection. The differential diagnosis should concern, first of all, Crohn's disease, colitis of various etiologies, as well as malignant neoplasm of the colon. An annual examination of bilirubin and cholestasis, which may exclude primary sclerosing cholangitis, PSC, is important for the prognosis of the disease (Kucharzik et al. 2020).

Ultrasound is a relatively new method in intestinal imaging. The advantage of this study is that no special preparation is required from the patient. Only a 6-hour fast is recommended to reduce the amount of air in the intestine. Its presence may interfere with the test results. Image quality can be improved by oral administration of polyethylene glycol, but it is not necessary. The transabdominal ultrasound begins with an indicative B-mode scan using a convex probe. For detailed examination of the intestines, high-frequency linear probes are used from 7.5 to 14 MHz. Five layers should be visible in the wall of a healthy intestine, the first and fifth are the reflection of ultrasound. An increase or decrease in the number of layers may indicate pathology. In order to exclude any irregularities, the thickness of the remaining 3 layers should be measured. The typical results of ultrasound in active ulcerative colitis are increased thickness of the intestinal mucosa and submucosa (BWT), greater than 4.0 mm in adults and 3.0 mm in children. Other results of ultrasound examination in UC patients include: fibrous-fat proliferation of the mesentery, enlargement of the mesenteric lymph nodes, loss of haustration, and sometimes increased echogenicity of the colonic submucosa. With the help of ultrasound Doppler, blood flow through the vessels can be detected. Due to the motor artifacts, the patient should hold his breath during such an examination. A typical result in patients with active ulcerative colitis is an increased color Doppler signal, which indicates an increased blood flow (Kucharzik et al. 2020).

X-Ray

When it is not necessary to obtain tissue for examination, x-ray examinations may be an alternative to endoscopic examination of the intestines in ulcerative colitis. This imaging can help determine the extent and severity of the disease and identify complications. For example, colon enlargement or bowel perforation. Dilation is usually most noticeable in the transverse colon. In addition, other features of active inflammation can be detected, such as colonic fluid air concentration and loss of colon haustration. In the severe course of the disease, the picture of the intestine is irregular, lumpy, with mucous islands. The extent of the disease can also be seen on X-rays as the inflamed region of the colon will contain less or no stool. A special type of X-ray examination is a barium enema examination to assess the condition of the colon. The image obtained with the use of double contrast with the use of barium, in which the intestinal mucosa is covered with a layer of the element, and the light remains distended with air, makes it possible to determine the length of the colon stricture and its diameter. This method can also detect ulcerations and erosions in the intestine. Such imaging can detect 15-20% of severe UC, which is manifested by a lumpy ileocecal valve with a granular appearance at the end of the ileum (Sandborn et al. 2020).

Endoscopy plays a fundamental role in the diagnosis of ulcerative colitis. It includes, among others. Colonoscopy, flexible sigmoidoscopy or endoscopic ultrasound. Endoscopic examination makes it possible to distinguish UC from Crohn's disease or other diseases such as drug-induced colitis, infectious colitis or ischemic colitis. In addition, it allows you to monitor the condition of patients after treatment or surgery. In patients with suspected UC, colonoscopy with ileoscopy should be performed to visualize the colon and end of the ileum. A biopsy should also be performed. The material should be collected from five different locations, both from the affected and normal-looking mucosa. It is recommended to re-endoscopy 3-6 months after starting UC treatment. Moreover, such patients are at high risk of developing colorectal cancer, hence it is suggested to perform endoscopy every 1 year (Seyedian et al. 2019).

Magnetic resonance imaging (MRI) is a non-invasive method that allows you to obtain images from inside the examined object. Classic MRI with the use of diffusion technique without prior preparation of the intestine makes it possible to distinguish inflammatory segments of the intestine from those with normal mucosa, which is necessary in the diagnosis of ulcerative colitis. MRI allows you to visualize the entire colon. The advantage of magnetic resonance imaging is the speed of the examination, as well as the lack of exposure of the patient to harmful ionizing radiation (Seyedian et al. 2019).

There are many methods for diagnosing ulcerative colitis. They are based on the variety of pathological changes observed in the patient's body. UC is a non-specific inflammatory bowel disease, and therefore the correct diagnosis of this disease should not be made on the basis of a single diagnostic test.

References

Feuerstein JD, Isaacs KL, Schneider Y, Siddique SM, Falck-Ytter Y, Singh S; AGA Institute Clinical Guidelines Committee. AGA Clinical Practice Guidelines on the

Management of Moderate to Severe Ulcerative Colitis. *Gastroenterology.* 2020; 158(5): 1450-1461.

Flynn S, Eisenstein S. Inflammatory Bowel Disease Presentation and Diagnosis. *Surg Clin North Am.* 2019; 99(6): 1051-1062.

Kucharzik T, Koletzko S, Kannengiesser K, Dignass A. Ulcerative Colitis-Diagnostic and Therapeutic Algorithms. *Dtsch Arztebl Int.* 2020; 117(33-34): 564-574.

Sandborn WJ, Baert F, Danese S, Krznarić Ž, Kobayashi T, Yao X, Chen J, Rosario M, Bhatia S, Kisfalvi K, D'Haens G, Vermeire S. Efficacy and Safety of Vedolizumab Subcutaneous Formulation in a Randomized Trial of Patients With Ulcerative Colitis. *Gastroenterology.* 2020; 158(3): 562-572.e12.

Seyedian SS, Nokhostin F, Malamir MD. A review of the diagnosis, prevention, and treatment methods of inflammatory bowel disease. *J Med Life.* 2019;12(2): 113-122.

Chapter 24

Barley Enzymes

Dominika Leś, David Aebisher and Dorota Bartusik-Aebisher[*]

Medical College of the University of Rzeszów, Rzeszów, Poland

Abstract

Enzymes are multi-molecular, mostly protein catalysts that improve specific chemical reactions by lowering their activation energy. Almost all chemical reactions related to the functioning of living organisms, including viruses, require the participation of enzymes to achieve sufficient efficiency.

Keywords: enzymes, barley enzymes

Enzymes are highly substrate specific and therefore a given enzyme catalyzes only a few of the possible reactions for a given substrate. In this way, enzymes determine metabolic and biochemical processes related to the functioning of living organisms.

Amylolytic enzymes, amylases, diastases are a group of enzymes classified as hydrolases that break down starch and other polysaccharides. They are found in saliva (salivary amylase) and in pancreatic juice (pancreatic amylase). Amylases are also synthesized in the fruits of many plants during ripening, making them sweeter, and during the germination of cereal grains. Grain amylase is important in the production of malt (Guerra, et al., 2009).

---

[*] Corresponding Author's Email: dbartusikaebisher@ur.edu.pl.

In: The Biochemical Guide to Enzymes
Editors: David Aebisher and Dorota Bartusik-Aebisher
ISBN: 979-8-88697-410-2
© 2023 Nova Science Publishers, Inc.

In nature, there are 3 types of amylase: α (EC 3.2.1.1), β (EC 3.2.1.2) and γ (EC 3.2.1.3), while in humans and other animals only α and γ amylases. The presence of 8 isoenzymes derived from: mucosa of the small intestine: P2; salivary glands: S1, S2, S3; pancreas: P1, P2, P3; mammary glands: P2, S1, S2; ovarian and testicular cells: O2, O1. The amylase test is simple to perform and is the primary test for pancreatitis. Increased amylase levelsserum levels are also found in the case of: some malignant neoplasms (bronchial, thyroid, liver cancer), salivary gland injuries, renal failure, alcoholism, peritonitis, mumps.

Amylases are many of the enzymes used to hydrolyze starch into simpler sugars. Plant-based amylases have long been used in the food industry, especially in the brewing and spirit industries. Currently, the most commonly used in the food industry are amylases obtained with the use of microorganisms (Stanca, et al., 2016). The amylases that cleave the α - 1,4 - glycosidic bonds of starch include: α - amylase, β - amylase, exo - α - amylase. The α - 1,6 - glycosidic bonds are cleaved by isoamylase and pullulanase. Glucoamylase and some pullulanases can cleave α - 1,4 – bonds and 1,6 - glycosidic. Considering the position of the cleaved bond in the starch chain, we distinguish between endoamylases, e.g., α-amylases, amylases which abolish cyclodectrinase branching, and exoamylases, i.e., glucoamylases, β-amylases, exo - α - amylases. Taking into account the technological point of view, there are liquefying amylases - α - amylases, pullulanases, cyclodextrinases and saccharifying - some α - amylases, β - amylases, glucoamylases.

Research on the enzymatic degradation of starch has been going on for a long time.

In the starch industry, various enzymes or their mixtures can be used, which makes it easier to obtain products that meet the specific requirements of customers. The end products of starch hydrolysis are different and dependent on the enzyme used. The susceptibility of starch to the action of amylolytic enzymes depends on its origin. It was noted that corn and soy starch are hydrolyzed faster than potato starch. Thanks to their properties, amylolytic enzymes are widely used, among others. in the brewing industry, in the process of mashing malt to liquefy and saccharify starch in order to prepare it for fermentation by yeast (De Schepper, et al., 2022). Due to the use of amylolytic enzymes, the starch germination temperature can be lowered from approx. 75-80°C to a temperature of 70°C. Starch saccharification occurs as a result of the interaction of "sugar" β - amylase and "decoding" α - amylase, which is strongly influenced by environmental pH and temperature (Duke, et al., 2013). It should be remembered that the final effect of enzymes is influenced by many technological factors, including temperature and pH. Amylases of microbial origin are used in agricultural distilling in the saccharification of potato and cereal starch. They make it possible to partially or completely eliminate the malt.

The use of amylase of mold origin enables the replacement of malt by 50-60%, while the use of thermostable, bacterial α-amylase and amyloglucosidase (glucoamylase) of fungal origin allows for the complete elimination of malt. The dose of enzymes in terms of starch is 0.1-0.5% α - amylase and 0.15-0.4% amyloglucosidase. The use of amylolytic preparations in baking increases the reserve of fermented sugars used by yeast, which in turn increases the fluffiness of the dough. Amylolytic enzymes are also used in the cereal industry (for the production of crystal glucose, the so-called cereal syrup, for the production of dextrins). In the potato and starch industry for the production of various preparations, such as thickeners, additives for sauces, puddings, and nutrients for children. These preparations are often used in the production of glucose and syrups for the confectionery industry (for the production of candies and halva), fruit and vegetable (for the production of jams and marmalades), and also as a sweetener in the baking and brewing industries.) or in confectionery for the recovery of sugar from confectionery waste. Depending on the microorganism used for the production of α - amylase, the obtained enzymes differ in the location of the hydrolyzed α - 1,4 - glycosidic bond in starch and in the optimal hydrolysis conditions, i.e., optimal temperature and acidity. Thus, the end products of starch hydrolysis may be different depending on the origin of the α-amylase. This enzyme, synthesized by bacteria, can liquefy or liquefy and saccharify starch.

As a result of the starch liquefaction enzyme, we obtain a syrup with a DE of 30 (dextrose equivalent, i.e., the number of reducing sugars per unit of dry substance of the product), containing glucose, maltose, maltotriose, significant amounts of oligosaccharides containing 6 glucose residues and borderline dextrins. As a result of the action of saccharifying enzymes on starch, particles containing 1-6 glucose residues and α - borderline dextrins are obtained. As a result of starch hydrolysis, a syrup with a DE 50 is obtained. Due to the α - amylases synthesized by starch fungi, a syrup with a DE 30-80 can be obtained, containing glucose, maltose, oligosaccharides containing 3-6 glucose residues and α - dextrins border (Duke, et al., 2013). Enzymes are auxiliary substances that are added in small amounts and are broken down during production, and sometimes trace amounts can be found in the food produced. In the food industry, amylases are very often used in baking. Their activity is primarily influenced by temperature.

References

De Schepper, C. F., Buvé, C., Van Loey, A. M., Courtin, C. M. A kinetic study on the thermal inactivation of balery malt α-amylase and β-amylase during the mashing process. *Food Research International*. 2022, 157.

Duke, S. H., Vinje, M. A., Henson, C. A. Tracking amulolytic enzyme acitivities during congress mashing with north american barley cultivars: Comparison of patterns of activity and beta-Amylases with differing B*my*1 Intron III alleles and correlations of amylolytic enzyme activities. *Cerevisia*. 2013, 38, 2, 51-52.

Duke, S. H., Vinje, M. A., Henson, C. A. Comparisons of amylolytic enzyme activities and beta-Amylases with differing B*my*1 Intron III alleles to sugar production during congress mashing with north american barley cultivars. *Cerevisia*. 2013, 38, 2, 58.

Guerra, N. P., Torrado-Agrasar, A., López-Macías, C., Martínez-Carballo, E., García-Falcón, S., Simal-Gándara, J., Pastrana-Castro, L. M. 10 – Use of amylolytic enzymes in brewing. *Beer in Health and Disease Prevention*. 2009, 113-126.

Stanca, A. M., Gianinetti, A., Rizza, F., Terzi, V. Barley: An overview of a versatile cereal grain with many food and feed uses. *Reference Module in Food Science*. 2016.

Index

A

acetaldehyde dehydrogenase, ix, 13, 14, 16, 17, 32, 33
acetylcholine, 77, 78, 79, 80
acid, x, xi, 1, 8, 13, 20, 23, 26, 28, 41, 43, 53, 54, 55, 56, 61, 62, 63, 73, 79, 83, 85, 92, 99, 100, 101, 102, 112, 114, 115
activation energy, x, xii, 31, 123
active site, x, 10, 14, 21, 22, 50, 54, 57, 66, 73, 86, 89, 100
adenine, 20, 33, 34
aerobic, xi, 65, 67
aesthetic, xii, 36, 99, 101, 109
age, 29, 43, 60, 63, 79, 92, 106, 113
aging process, xi, 105, 108
ALA synthase/aminolevulinic acid synthase (ALAS), x, 53, 54, 55, 56
amino acids, 10, 14, 21, 41, 48, 49, 61, 63, 90, 100
aminolevulinic acid synthase, x, 53, 54, 55, 56
amylase, 47, 49, 123, 124, 125, 126
anaerobic, xi, 6, 65, 67, 72, 74, 75
anemia, 38, 53, 55, 118
apoptosis, x, 2, 7, 19, 22, 23, 36, 42, 86, 110
archaea, xi, 65, 71
atoms, 10, 33, 34, 58, 61, 73
ATP, ix, 6, 9, 11, 28, 29, 96
attachment, 21, 22, 74
autosomal recessive, 26, 28, 61

B

bacteria, xi, 10, 14, 65, 67, 71, 83, 85, 92, 100, 101, 106, 107, 125
beta-galactosidase, xi, 105, 106, 107, 108
biochemical, ix, x, 1, 2, 3, 7, 29, 41, 79, 87, 97, 123
biochemistry, xi, 8, 9, 13, 19, 23, 31, 35, 53, 65, 67, 69, 71, 77, 83, 89, 95, 97, 99, 105, 109, 117
biomarker, xii, 96, 97, 105, 107, 108
biopsy, 50, 111, 118, 120
biosynthesis, 10, 53, 55, 74
blood, 20, 32, 42, 49, 50, 53, 54, 58, 61, 62, 63, 101, 110, 118, 119
bonding, xii, 68, 105
bonds, x, xi, 2, 47, 48, 49, 61, 90, 99, 101, 124
bowel, xii, 28, 92, 117, 118, 119, 120, 121
brain, 7, 22, 32, 55, 58, 59, 60, 63, 79, 96, 97, 100
breakdown, 16, 26, 28, 90, 106

C

cancer, ix, 1, 2, 3, 4, 7, 13, 17, 23, 32, 34, 36, 38, 42, 43, 45, 86, 87, 95, 96, 107, 108, 111, 116, 120, 124
cancer cells, 32, 36, 38, 107
carbohydrates, x, 3, 28, 29, 41, 47, 48, 49, 59, 108
carbon, xi, 6, 9, 29, 33, 61, 71, 75, 92, 101, 106
carbon dioxide, 29, 92, 106
carcinoma, xii, 97, 105, 111

catalyst, x, 31, 67, 68
catalytic activity, 14, 58, 61, 79
catalyzes, xi, 11, 27, 41, 74, 96, 99, 123
cell, x, xi, 2, 3, 5, 6, 7, 8, 10, 11, 16, 17, 19, 20, 21, 22, 23, 32, 34, 35, 36, 37, 38, 41, 42, 43, 44, 45, 50, 53, 54, 55, 57, 58, 59, 60, 67, 79, 80, 84, 86, 87, 89, 90, 91, 95, 96, 97, 99, 101, 103, 105, 107, 108, 109, 110, 111, 114, 115, 124
cell aging, xi, 86, 105
cell division, 6, 35, 36
central nervous system, xi, 62, 77, 86
chemical, ix, xi, xii, 2, 10, 11, 22, 61, 67, 100, 108, 112, 123
chemical reactions, xii, 11, 123
childhood, 3, 4, 55
children, 29, 113, 119, 125
choline acetyltransferase (ChAT), xi, 77, 78, 79, 80
cholinergic nerve cells, xi, 77
chromosome, ix, xi, 5, 6, 20, 35, 48, 58, 59, 62, 89, 90, 100, 106
classes, ix, 9, 28, 85
cleavage, 34, 106, 107
coenzyme, 21, 31, 33, 34
colitis, xii, 117, 118, 119, 120
colon, xii, 107, 117, 118, 119, 120
colorectal cancer, 107, 108, 120
colorectal tumor imaging, xii, 105
complex carbohydrates, x, 29, 47
composition, x, 31, 100
compounds, 1, 14, 26, 36, 39, 49, 50, 83, 86, 96, 101, 108
consumption, 4, 28, 59, 60, 67, 91, 92, 108
cosmetic, xii, 36, 109, 111, 115
cyanobacteria, xi, 65, 67, 68
cysteine, 6, 7, 14, 20, 21, 22, 49, 73, 83, 85, 86
cytochrome, 10, 11, 54
cytoplasm, 5, 6, 20, 55, 90, 95, 96
cytoskeleton, x, 6, 19, 23, 85

D

decarboxylation, ix, 9, 56

decomposition, x, 25, 102, 103
deficiency, 2, 3, 4, 16, 26, 28, 29, 49, 50, 53, 55, 64, 75, 92, 100, 102, 118
degradation, xi, 7, 15, 60, 99, 106, 107, 124
dehydrogenase, v, vi, ix, x, 1, 2, 3, 4, 5, 6, 7, 8, 13, 14, 16, 17, 19, 20, 22, 23, 31, 32, 33, 34, 41, 42, 43, 44, 45, 72, 73
dermatological therapies, xii, 109
dermatologist, 112, 113, 114, 115
detection, xii, 62, 105, 107, 108
detoxification, 13, 16, 54
diabetes, 49, 50, 58, 60
diagnosis, xii, 3, 29, 50, 64, 105, 107, 108, 116, 117, 118, 120, 121
diarrhea, xii, 29, 48, 50, 91, 106, 117
diet, x, 4, 29, 47, 62, 63
digestion, xi, 28, 48, 49, 50, 90, 93, 105, 106
dimethyl sulfoxide reductases (DMSO), xi, 71, 72, 73, 74, 75
disease, ix, xii, 1, 2, 3, 4, 7, 8, 12, 13, 14, 17, 19, 23, 28, 32, 37, 38, 39, 42, 43, 44, 45, 48, 49, 50, 51, 53, 55, 58, 60, 61, 62, 63, 78, 79, 80, 86, 87, 92, 95, 96, 97, 98, 100, 102, 107, 108, 111, 115, 117, 118, 119, 120, 121, 126
disorders, xi, 2, 3, 4, 28, 29, 39, 42, 43, 45, 51, 58, 62, 78, 79, 80, 105
distribution, 15, 68, 79
DMSO reductase, 71, 72, 73, 74, 75
DNA, x, 3, 6, 7, 8, 23, 35, 38, 101
doctors, ix, 1, 3, 102, 112, 113, 114
drugs, 7, 49, 54, 55, 87, 101, 102, 114

E

electron, 2, 10, 16, 31, 33, 34, 67, 69, 74, 75
emission, 9, 10, 11, 16, 67, 107
encoding, xi, 33, 49, 89, 90, 91, 92, 100, 106
endothelial cells, 58, 59, 86
energy, x, xii, 2, 3, 6, 10, 22, 25, 29, 31, 41, 50, 60, 63, 90, 96, 110, 123

Index

engineering, 10, 11, 12
enterocyte cell membrane, xi, 89, 90
environment, 9, 14, 21, 25, 54, 67, 68
enzymes, vii, ix, x, xi, xii, 2, 5, 6, 7, 9, 10, 11, 13, 14, 16, 19, 20, 21, 22, 23, 25, 26, 28, 29, 31, 32, 33, 34, 35, 36, 37, 38, 39, 41, 42, 43, 47, 48, 49, 50, 53, 54, 55, 57, 58, 59, 61, 62, 63, 65, 66, 67, 68, 71, 72, 73, 74, 75, 77, 78, 79, 80, 83, 85, 86, 89, 90, 91, 92, 95, 96, 99, 100, 101, 105, 106, 107, 108, 117, 123, 124, 125, 126
erythrocytes, 54, 55, 95
Escherichia coli, xi, 72, 73, 83
ethanol, 13, 15, 32
etiology, xii, 28, 45, 117
eukaryotic, xi, 5, 35, 36, 54, 65
exposure, 42, 109, 111, 120

F

fatty acids, 31, 41, 78, 106
fertilization, 67, 101, 102
fixation, 66, 67, 68
fluid, 58, 95, 99, 102, 114, 119
fluorescent, xii, 10, 105, 107, 108
food, xii, 30, 50, 67, 102, 105, 106, 107, 108, 124, 125, 126
food industry, xii, 105, 108, 124, 126
formation, x, 6, 10, 11, 17, 19, 23, 33, 34, 47, 54, 55, 61, 62, 74, 78, 85, 89, 90, 107, 111
fructose, 25, 27, 28, 29, 34
fungi, 10, 106, 125

G

gene, ix, x, xi, 5, 8, 10, 11, 13, 16, 19, 20, 29, 33, 36, 38, 43, 45, 48, 49, 58, 59, 60, 61, 62, 74, 78, 79, 84, 86, 89, 90, 91, 92, 100, 102, 106, 107
genetic information, 6, 36, 106
genome, 6, 38, 106
germination, 123, 124
gland, 58, 59, 60, 124
glucoamylase, 28, 30, 39, 47, 48, 49, 50, 51, 125

glucose, 2, 3, 25, 27, 28, 29, 32, 39, 48, 49, 50, 60, 89, 96, 105, 125
glutaredoxins (Grxs), vi, xi, 83, 84, 85, 86, 87
glutathione, xi, 83, 85
glyceraldehyde-3-phosphate dehydrogenase (GAPDH), ix, 5, 6, 7, 8, 19, 20, 21, 22, 23
glycolysis, ix, 2, 5, 6, 7, 19, 20, 21, 22, 23, 85
glycolytic enzyme enolase, xi, 95
growth, 58, 63, 67, 96, 107

H

half-lives, x, 10, 11, 57, 97
health, xii, 92, 109
heme, x, 53, 54, 55, 110
hemoglobin, x, 53, 54, 55, 110
heterotrophs, xi, 65, 67
homeostasis, 42, 44, 85, 87
hormones, 57, 58, 59, 60
human, ix, x, xi, 5, 8, 13, 14, 15, 16, 20, 23, 25, 35, 36, 37, 43, 55, 57, 63, 83, 86, 99, 100, 103, 105, 106, 107
human body, ix, x, 5, 13, 15, 25, 36, 55, 57, 99, 103, 106
hyaluronic acid, xi, 99, 100, 101, 102
hyaluronidase, vii, xi, 99, 100, 101, 102, 103
hydrogen, 2, 10, 21, 26, 31, 33, 34, 74, 75, 83, 92, 106
hydrolysis, ix, xi, 9, 25, 27, 28, 47, 48, 49, 61, 100, 106, 123, 124, 125
hypoxia, 2, 55, 59

I

identification, 54, 68, 107
ileum, xii, 117, 120
in vitro, 6, 11, 38, 39, 67, 106, 108
in vivo, 6, 11, 38, 106, 107, 108
incidence, 14, 16, 43, 62, 91, 118
industry, xii, 36, 105, 108, 124, 125, 126
inflammation, 92, 96, 97, 114, 118, 119

inflammatory bowel diseases, xii, 92, 117, 118, 120, 121
inhibition, 2, 22, 38, 39, 43, 49, 50, 55
injuries, xii, 42, 95, 96, 97, 109, 124
insertion, 37, 68, 74
insulin, 50, 60, 74
intestine, xi, xii, 25, 26, 28, 29, 47, 49, 63, 89, 92, 106, 117, 119, 120, 124
intracellular, xi, 29, 55, 83
ions, xi, 2, 26, 47, 61, 65
iron, x, 19, 55, 61, 68, 85, 87, 118
iron transport, x, 19, 85, 87
isoforms, x, 2, 54, 55, 57, 68, 83, 85

K

Krebs cycle, 1, 31, 41, 43, 44

L

lactase, xi, 89, 90, 91, 92, 93, 106
lactase phlorozin hydrolase (LPH), xi, 89, 90, 91
lactose, x, xi, 47, 89, 91, 92, 93, 105, 106, 107, 108
lactose intolerance, 91, 93, 108
large intestine, xii, 49, 106, 117
lead, 2, 7, 16, 55, 61, 62, 63
lesions, xii, 14, 109, 111, 112, 113
light, xii, 9, 10, 11, 109, 110, 112, 113, 115, 120
liver, xi, 13, 17, 32, 53, 55, 58, 59, 60, 61, 63, 108, 124
L-phenylalanine, xi, 61
L-tyrosine, xi, 61
luciferase, v, ix, 9, 10, 11, 12
luciferin, ix, 9, 10, 11
lumen, 25, 28, 90, 106

M

macrophages, 4, 16, 96
maltase, 30, 39, 47, 48, 50, 51
maltose, 28, 29, 48, 125
mammals, xi, 5, 16, 85, 89, 91, 92, 100, 101

mass, 2, 5, 20, 57, 72
matrix, 16, 54, 84, 99
medical, 2, 29, 109
medicine, 36, 75, 99, 101, 102, 108
metabolic, x, 1, 2, 3, 4, 16, 28, 41, 43, 44, 51, 55, 59, 60, 69, 85, 123
metabolism, 2, 3, 4, 6, 14, 31, 32, 41, 44, 55, 57, 58, 59, 60, 67, 71, 72, 83, 110
metalloenzymes, xi, 65, 95
metastasis, 3, 43, 97
mitochondria, x, 1, 2, 3, 5, 14, 16, 20, 42, 43, 44, 45, 53, 54, 84, 86
molecular weight, xi, 2, 20, 36, 78, 89, 90
molecules, x, 2, 3, 4, 6, 10, 33, 38, 47, 49, 50, 55, 60, 68, 73, 74, 85, 86, 90, 100, 110
molybdenum, 68, 71, 72, 73, 74, 75
monosaccharide, x, 4, 27, 28, 47, 89, 92
mucosa, xii, 50, 117, 119, 120, 124
multi-molecular, xii, 102, 123
mutant, xi, 14, 83
mutation, ix, 13, 16, 26, 28, 42, 43, 49, 55, 58, 61, 62, 64, 79, 91, 92, 100, 102, 106, 108

N

NAD, 6, 20, 21, 22, 31, 33, 34
NADH, 6, 44, 96
NADPH, xi, 83, 84
neoplastic diseases, ix, 1, 44, 45, 96
nerve, xi, 22, 75, 77, 80, 96, 97
nervous system, xi, 59, 62, 77, 79, 80, 86
neurodegenerative diseases, 7, 8, 19, 23, 42, 44, 78, 86
Neuron Specific Enolase (NSE), vi, xi, 95, 96, 97
neurons, 59, 77, 78, 79, 80, 86, 95, 96
nitrogen, 65, 66, 67, 68, 71, 75
nitrogen fixation, 66, 67, 68
nitrogenase, 66, 67, 68, 69
nitrogenases (nitrases), xi, 65, 66, 68
nucleotides, 33, 35, 36, 37
nucleus, 5, 6, 16, 20, 22, 37, 79, 86, 96
nutrients, 48, 50, 63, 68, 125

O

obesity, 14, 49, 60
organism, x, xi, xii, 6, 7, 10, 11, 16, 35, 37, 38, 41, 42, 53, 60, 65, 68, 71, 72, 96, 99, 102, 105, 107, 123
organs, 13, 32, 34, 42, 55, 60
ovarian cancer, xii, 105, 107, 108
oxaglutarate dehydrogenase, x, 41, 42, 44
oxidation, ix, 1, 2, 5, 7, 9, 10, 11, 16, 21, 31, 32, 33, 34, 45, 56, 72, 74, 84, 85, 86
oxidative stress, 7, 32, 56, 85, 86
oxygen, ix, x, 2, 6, 9, 10, 11, 16, 33, 42, 44, 53, 55, 59, 67, 71, 73, 74, 75, 110

P

pathology, 8, 23, 107, 112, 119
pathway, 1, 7, 36, 41, 42, 44, 48, 50, 68, 79, 80, 85, 110
pathways, 7, 41, 44, 68, 79, 85
pH, 9, 25, 29, 32, 34, 100, 124
phenotype, 62, 80, 92
phenylalanine hydroxylase, xi, 61, 62, 63, 64
phenylketonuria, 61, 62, 63, 64
phosphate, ix, 5, 6, 7, 8, 9, 19, 20, 21, 22, 23, 32, 33, 54
phosphorylation, x, 2, 15, 19, 21, 74, 78, 79
photodynamic therapy (PDT), xii, 109, 110, 111, 112, 113, 114, 115, 116
photosensitivity, 55, 113, 115
plants, 67, 106, 123
polymerase, x, 35, 37
polymorphism, 14, 60, 79, 106
polypeptide, 2, 20, 90
polysaccharide, 101, 123
population, 14, 28, 60, 62, 92
porphyrin, x, 53, 54, 110
pregnancy, 43, 58, 63
preparation, 102, 119, 120
prevention, 14, 63, 121
probability, 60, 62, 118
prognosis, 3, 95, 96, 119
prokaryotes, xi, 71, 72, 74, 75

promoter, 10, 78, 91, 106
protection, 14, 86, 115
protein, ix, x, xi, xii, 5, 6, 7, 8, 9, 10, 11, 12, 14, 15, 19, 20, 21, 22, 23, 25, 26, 28, 29, 31, 33, 37, 41, 42, 48, 51, 53, 54, 57, 58, 59, 60, 61, 63, 65, 66, 67, 69, 72, 73, 74, 75, 77, 78, 79, 85, 86, 87, 89, 90, 91, 95, 96, 99, 102, 108, 110, 118, 123
pyruvate dehydrogenase, 1, 2, 3, 4

R

reaction, ix, x, xi, xii, 1, 6, 9, 10, 11, 12, 16, 17, 21, 23, 25, 26, 31, 32, 33, 34, 41, 45, 49, 54, 57, 58, 61, 71, 72, 74, 85, 86, 100, 102, 110, 114, 115, 123
reactive oxygen, 2, 16, 42, 43, 110
receptors, 58, 59, 80
recovery, 29, 85, 125
red blood cells, 20, 32, 53, 54, 110
reductases, xi, 71, 72, 73, 83
regeneration, 75, 92, 96
replication, 7, 35, 37, 38
research, xi, 2, 3, 4, 8, 14, 16, 22, 23, 26, 29, 36, 38, 42, 49, 51, 66, 67, 68, 71, 83, 85, 86, 97, 100, 101, 105, 107, 124, 126
researchers, 14, 42, 43, 75, 92
residues, 1, 2, 6, 14, 22, 54, 72, 73, 79, 100, 106, 110, 125
respiration, 29, 67, 72, 74, 75
response, 50, 83, 96, 97
reticulum, 7, 16, 57
ribonucleoprotein complex (RNP), x, 35, 38
risk, 33, 79, 112, 120
RNA, x, 19, 20, 35, 37, 56, 78

S

science, xii, 29, 87, 92, 108, 109
selenium, 58, 72, 73, 75
sensitivity, 11, 60, 107
side effects, 49, 102, 112, 113, 114, 115
skeletal muscle, 16, 58, 59
skin, xii, 3, 55, 58, 59, 102, 109, 110, 111, 112, 113, 115

skin conditions, xii, 109, 111
small intestine, xi, 25, 26, 28, 29, 47, 63, 89, 124
species, 2, 16, 42, 43, 66, 68, 110
stabilization, x, 19, 37, 72
starch, 26, 28, 29, 49, 50, 123, 124, 125
state, x, 31, 32, 59, 71, 77, 80, 97, 110
stress, 7, 32, 56, 83, 85, 86
structure, ix, x, 6, 7, 10, 11, 14, 20, 22, 29, 31, 32, 35, 37, 41, 42, 48, 50, 54, 58, 66, 67, 72, 73, 74, 75, 83, 85, 99, 100, 107
substrate, 1, 10, 11, 14, 32, 33, 34, 48, 49, 50, 54, 68, 75, 100, 107, 123
sucrase-isomaltase, 25, 28, 29, 30, 47, 50, 51
sucrose, v, 25, 27, 28, 29, 107
sugars, xii, 26, 29, 47, 48, 49, 50, 89, 105, 106, 108, 124, 125, 126
sulfur, 71, 73, 75, 85
symptoms, xii, 3, 26, 28, 29, 43, 50, 60, 62, 91, 92, 97, 115, 117, 118
syndrome, 28, 58, 60, 79, 115
synthesis, x, 1, 6, 10, 37, 53, 54, 55, 56, 77, 78, 79, 80, 83, 85, 106

T

target, 4, 34, 58, 59, 87, 97
techniques, 106, 108, 112
technology, 10, 54, 103, 108
telomerase, v, x, 35, 36, 37, 38
telomere, 6, 7, 37, 38
temperature, 9, 10, 11, 124, 125, 126
therapy, xii, 4, 48, 50, 62, 63, 109, 110, 111, 112, 113, 114, 115, 116
thioredoxins, 83, 85, 87

thyroid, 57, 58, 59, 60, 124
tissue, 2, 32, 58, 63, 96, 100, 102, 119
transcription, 2, 15, 42, 43, 58, 60, 78, 90
transformation, 21, 28, 41, 64, 78
translation, 11, 29, 90
translocation, 6, 15, 37
transport, x, xi, 2, 6, 7, 14, 19, 22, 47, 53, 77, 78, 85, 87
treatment, ix, xii, 1, 2, 3, 4, 29, 38, 42, 43, 44, 49, 50, 51, 63, 64, 79, 80, 97, 99, 109, 111, 112, 114, 115, 116, 117, 120, 121
tumor, xii, 2, 3, 58, 97, 102, 105, 109
type 2 diabetes, 50, 58, 60
type I deiodinase (D1), x, 57, 58, 60
type II deiodinase (D2), x, 57, 58, 60
type III deiodinase (D3), x, 57, 58, 59, 60, 80
tyrosine, xi, 2, 61, 63, 64

U

ulcerative colitis (UC), xii, 117, 118, 119, 120

V

vanadium, xi, 65, 68

W

water, xi, 29, 33, 48, 61, 74, 99, 100, 106, 114

Y

yeast, 38, 124, 125

Editors' Contact Information

David Aebisher, PhD, DSc
Professor
Medical College of the University of Rzeszów,
Rzeszów, Poland
Email: daebisher@ur.edu.pl

Dorota Bartusik-Aebisher, PhD, DSc
Professor
Medical College of the University of Rzeszów,
Rzeszów, Poland
Email: dbartusik-aebisher@ur.edu.pl